AF598047

Structural Health Monitoring Using Genetic Fuzzy Systems

Prashant M. Pawar • Ranjan Ganguli

Structural Health Monitoring Using Genetic Fuzzy Systems

Prashant M. Pawar
College of Engineering
Shri Vithal Education and Research Institute
Pandharpur
India
pawarpm@gmail.com

Ranjan Ganguli
Department of Aerospace Engineering
Indian Institute of Science
Bangalore
India
ganguli@aero.iisc.ernet.in

Additional material to this book can be downloaded from http://extras.springer.com.

ISBN 978-0-85729-906-2 e-ISBN 978-0-85729-907-9
DOI 10.1007/978-0-85729-907-9
Springer London Dordrecht Heidelberg New York

British Library Cataloguing in Publication Data
A catalogue record for this book is available from the British Library

Library of Congress Control Number: 2011933278

Cover design: VTeX UAB, Lithuania

Printed on acid-free paper

Springer is part of Springer Science+Business Media (www.springer.com)

Preface

Structural health monitoring (SHM) has emerged as an important research area in recent years because of its strong links with structural safety and the need to monitor and extend the lives of existing structures. SHM is an interdisciplinary field, combining elements of mechanics with those of information science and sensors and actuators. The practical importance of SHM is clear from the continuing failures which affect engineering structures such as bridges, aircraft, helicopters, and nuclear reactors. In many cases, a health monitoring system installed on the structure can detect and isolate the damage before it becomes catastrophic, thereby reducing the likelihood of failures. SHM systems can therefore reduce costs and save lives.

A key problem in SHM involves performing damage detection and isolation from a set of measured data. Typically, the measured data is contaminated with noise, and the number of measurements is limited. In model-based SHM, a mathematical model is used to develop simulated measured data for the damaged structure. Then, the simulated data is used to develop a pattern recognition approach which maps the damage location and size to the simulated data. Algorithms such as neural networks are often used to perform this pattern recognition task. However, neural networks tend to be black boxes which are difficult to understand. In this book, an alternative and powerful architecture, the genetic fuzzy system (GFS), is demonstrated for beams, composite tubes, and helicopter rotor blade health monitoring. A novel feature of this book is the focus on helicopter rotor health monitoring, as this represents a system of considerable complexity.

The fuzzy logic approaches addresses uncertainty directly through the linguistic fuzzifier and is very well suited to SHM because it gives linguistic outputs which can be used to guide prognostic action. The use of the genetic algorithm automates the development of the fuzzy system and makes the method easy to use for problems involving a large number of measurements and damage location sizes, which is typical of SHM. By demonstrating the use of the GFS as a series of progressively complicated structures, this book enables the reader to learn about this new and powerful approach to SHM. The book also provides some MATLAB code for the algorithms developed. This book will be useful for aerospace, civil, and mechanical engineers working in the area of SHM. It will also be useful for computer scientists and applied mathematicians interested in the application of GFSs to engineering problems.

Bangalore, India

Prashant M. Pawar
Ranjan Ganguli

Contents

1 Introduction . . . 1
1.1 Terms and Definitions Related to SHM . . . 2
1.2 SHM Approaches . . . 3
1.2.1 Model-Based SHM . . . 4
1.2.2 Modal-Based Methods . . . 8
1.2.3 Localized Methods . . . 9
1.3 Soft Computing Methods for Health Monitoring . . . 11
1.3.1 Neural Networks . . . 11
1.3.2 Genetic Algorithms . . . 15
1.3.3 Fuzzy Logic System . . . 16
1.3.4 Hybridized Soft Computing Methods . . . 18
1.4 Book Summary . . . 20
References . . . 21

2 Genetic Fuzzy System . . . 25
2.1 Fuzzy Logic System . . . 26
2.2 Genetic Algorithms . . . 31
2.2.1 Operations During a GA Process . . . 33
2.2.2 Performance Factors . . . 35
2.3 Genetic Fuzzy System . . . 36
2.4 Summary . . . 39
References . . . 39

3 Structural Health Monitoring of Beams . . . 41
3.1 Damage Modeling in 1D Beam . . . 42
3.2 Formulation of Genetic Fuzzy System . . . 45
3.3 Structural Health Monitoring for Uniform Beam . . . 47
3.3.1 Test with Noisy Data . . . 48
3.3.2 Test with Different Measurements . . . 51
3.3.3 Missing and Faulty Measurements . . . 52
3.3.4 Measurements with High Noise . . . 53

3.4 Structural Health Monitoring for Nonuniform Beam 54
3.4.1 Refined Output Set 57
3.5 SHM for BO105 Hingeless Helicopter Rotor Blade 60
3.5.1 Frequency-Based Damage Detection of Blade 60
3.5.2 Mode Shape-Based Damage Detection of Blade 63
3.6 Summary 65
References 66

4 Structural Health Monitoring of Composite Tubes 67
4.1 Matrix Cracking in Hollow Circular Cross Section 68
4.1.1 Effective Elastic Modulus 68
4.1.2 Matrix Crack Model ($[\pm\theta_m/90_n]$ Family of Composites) 69
4.2 Modal Analysis 72
4.3 Damage Detection in Composite Structure 75
4.3.1 Development of Genetic Fuzzy System 75
4.3.2 Testing of Genetic Fuzzy System 78
4.3.3 Analysis of Misclassification 81
4.4 Summary 82
References 83

5 Structural Health Monitoring of Composite Helicopter Rotor 85
5.1 Mathematical Model 86
5.1.1 Mathematical Model of Helicopter Rotor 86
5.1.2 Composite Rotor Blade 92
5.1.3 Progressive Damage Accumulation 93
5.2 Data Reduction 97
5.3 Life of the Structure 98
5.4 Behavior of Composite Rotor Blade 99
5.4.1 Effect on Cross-Sectional Stiffness 100
5.4.2 Effect on Static Response 101
5.5 Numerical Simulation of Measurement Deltas 103
5.5.1 Blade Tip Response 103
5.5.2 Blade Root Loads 105
5.5.3 Thresholds Based on Matrix Crack Saturation 108
5.5.4 Strains 108
5.6 Predicting Life Consumption 110
5.6.1 Development of Genetic Fuzzy System 111
5.6.2 Testing of Genetic Fuzzy System 115
5.6.3 Implementation of the SHM System 124
5.7 Summary 124
References 125

Appendix MATLAB Codes 127

Index 129

Chapter 1
Introduction

Structural health monitoring (SHM) is necessary for various aerospace, mechanical, and civil engineering applications for evaluating the fitness of a structure to perform its prescribed tasks. The structure may change its performance based on a gradual or sudden change in state, load conditions, or response mechanisms. The main objective of developing the SHM system is to enhance structural safety. However, SHM serves other economic benefits such as increased mission reliability, extended life of life-limited components, reduced tests, reduction in "down time," increased equipment reliability, customization of maintenance actions, and greater awareness of operating personnel, resulting in fewer accidents [1]. SHM also promises to help in reducing maintenance costs. For example, in the case of helicopters, maintenance costs account for about 24% of the direct operating costs [2]. Due to evolution in on-board information processing technology and avionics, comprehensive SHM systems are feasible, and flight data recorder systems mitigate the cost of installing the SHM systems [3]. Hence, SHM plays an important role in the case of aging structures (whose life span is increased due to budget restrictions) by promising an increase in safety and a reduction in the maintenance cost.

Some aerospace vehicles are equipped with a health and usage monitoring system (HUMS), which is a very good example of SHM. HUMS is highly recommended for helicopters, as these flight vehicles are prone to damage due to high vibration levels. The process of usage monitoring involves assessing the life consumption of critical component systems and structures by monitoring actual damage exposure. For predicting the remaining life of components, accurate prediction of the total life of a component and accurate measurement of its usage is needed. The simplest form of usage monitoring is time monitoring. The use of automated time monitoring not only relieves pilot and crew workload, it also removes errors associated with the rounding of flight hours. Another function of HUMS is automated recording in flight exceedance, thus removing the dependency upon the crew to perform this task manually. For example, if a predefined oil temperature, oil pressure, or shaft torque level is exceeded, then HUMS records the time and level of the exceedance for post flight evaluations. In addition, HUMS records the values of a set of flight parameters (such as altitude, airspeed, and outside temperature) at the time of the exceedance.

P.M. Pawar, R. Ganguli, *Structural Health Monitoring Using Genetic Fuzzy Systems*,
DOI 10.1007/978-0-85729-907-9_1, © Springer-Verlag London Limited 2011

These parameter levels give additional insight to the maintainer during a post flight analysis. The faults detected through exceedance monitoring cover the engine, drive train, rotor, and oil system. Based on the experience of the Eurocopter HUMS development of more than 10,000 flight hours supporting 70 systems world wide, actual achievements were compared with initial expectations in [4]. From these studies, it was observed that the implementation of HUMS has numerous benefits. In particular, the safety enhancement provided by HUMS is now unquestioned.

In this book, we will consider both SHM and HUMS systems. HUMS goes beyond SHM in terms of sensor measurements; however, the underlying philosophy is the same: to use measurements to monitor system health. Several different structures will be considered with different damage modeling approaches. These applications are relevant to aerospace, civil, and mechanical engineering.

1.1 Terms and Definitions Related to SHM

Over the last two decades, SHM technology has shown enormous development. Various SHM-related terms have evolved which are defined below [5]. These terms occur frequently in this book and in the SHM literature.

Damage: Damage is defined as "a deficiency or deterioration in the strength of a structure, caused by external loads, environmental conditions, or human errors." Physically, damage may be visible as a crack, delamination, debonding, reduction in thickness/cross section, or exfoliation. The damage may not be easily visible to the naked eye. For example, composite structures are susceptible to barely visible impact damage (BVID). Operational structures are designed to operate with some amount of damage; i.e., they are damage tolerant. The term "damage" therefore carries a very different meaning compared to the term "failure." However, early damage detection can prevent failure of structures just as early detection of heart disease and cancer can prevent premature death in a human. Once the damage is detected, appropriate action needs to be taken to repair the structure.

Detection: Detection is defined as "the identification of existence of an anomalous condition in a system." For instance, by performing rotor track and balance in a helicopter rotor, the maintainer may identify an abnormally high fuselage vibration in the vertical direction at the one-per-rev ($1\ \Omega$) frequency, where Ω is the rotation speed. This measurement which can be used for damage detection is called a damage indicator. Detection includes active and passive sensing, signal processing, feature extraction, and data fusion.

Diagnostics: Diagnostics is defined as "the identification of the cause of the anomalous condition." For instance, the cause of the previously mentioned one-per-rev vibration in a helicopter rotor may be identified as a blade tab error on one blade of the rotor. Diagnosis includes the mapping of damage indicators to damage.

Prognostics: Prognostics is "the prediction of the amount of safe life left in a system." For instance, prognostication of a bridge would predict time until failure of the bridge. Currently, HUMS prognosticates only by measuring usage. To be truly predictive, a connection between the evolution of damage indicators and remaining useful life of the structure must be found. Thus, life prediction algorithms can play an important role in prognostics.

HUMS: HUMS is "a system and a process of avionics equipment, ground support equipment and procedures and putting of these into productive use." HUMS applied to aircraft will focus on onboard monitoring associated to online or ground data processing to create diagnostic and prognostic decisions. The HUMS are designed to improve the safe operation and reduce operating costs of aircraft.

Monitoring: HUMS employs sensors and computer-based processes on the aircraft along with associated computers and processes on the ground to continuously observe the damage indicators of aircraft members. These damage indicators may be as simple as engine or rotor speed or as complex as a fast Fourier transform (FFT) or wavelet transform of non-dimensional parameters associated with gears or bearings.

Usage (Monitoring): Usage is "a process which assesses the life consumption of critical components, systems, and structures by monitoring actual damage exposure."

Failure: Failure "refers to any action leading to an inability on the part of a structure or a machine to function in the intended manner." Fracture, permanent deformation, buckling, divergence, flutter, ground resonance, and even excessive linear elastic deformation may be regarded as modes of failure of structures. Failure results when a particular type of damage becomes too large, thereby seriously impairing the safety and/or the functioning of the structure.

Structural health monitoring (SHM): SHM "is defined as the acquisition, validation and analysis of a technical data to facilitate life cycle management decisions." SHM involves creation of a reliable system with the ability to detect and interpret "changes" in a structure due to damage or normal operations. SHM is highly multidisciplinary, drawing elements from the mechanical, information, and electrical sciences.

1.2 SHM Approaches

Traditional SHM methods were based on manual feedback such as abnormalities reported by means of warning, caution, advisory, and/or maintenance panels. The abnormalities were considered based on exceedance monitoring of various measurements such as vibration, temperature, pressure, strain, frequency, acceleration, displacement, and load. Exceedance monitoring involves the operations performed by the operators manually at various critical locations and for any critical components of the system to monitor that measurements do not exceed predefined thresholds. For example, in the transmission system, gear wear can be detected by post operation oil analysis using chip detectors [6]. Such SHM systems involve thresholding of sensor measurements and little information processing. Most of the effort here is on the development of better oil debris monitoring sensors. In turbine blades, a significant level of low cycle fatigue damage can be thresholded by monitoring natural frequencies [7]. Here, operational vibration data must be signal processed to extract online frequencies. Exceedance monitoring depends strongly on the quality of the trends in the measured data, and signal processing algorithms are used to remove noise and outliers from the signals [8]. In some systems, physics-based modeling capability is limited due to the complexity of the system. In these cases, exceedance monitoring may be the only feasible approach for health monitoring. The design of an SHM system depends on the type of structure, operating environment of the system, level of damage prediction accuracy, type of measurement signals available, and allowable cost of SHM. Based on these parameters, an SHM system can be developed using various approaches. These are summarized below with a few examples.

1.2.1 Model-Based SHM

In model-based SHM, damage is identified from changes in the simulated measurements from the structural model. Figure 1.1 represents a typical SHM, which

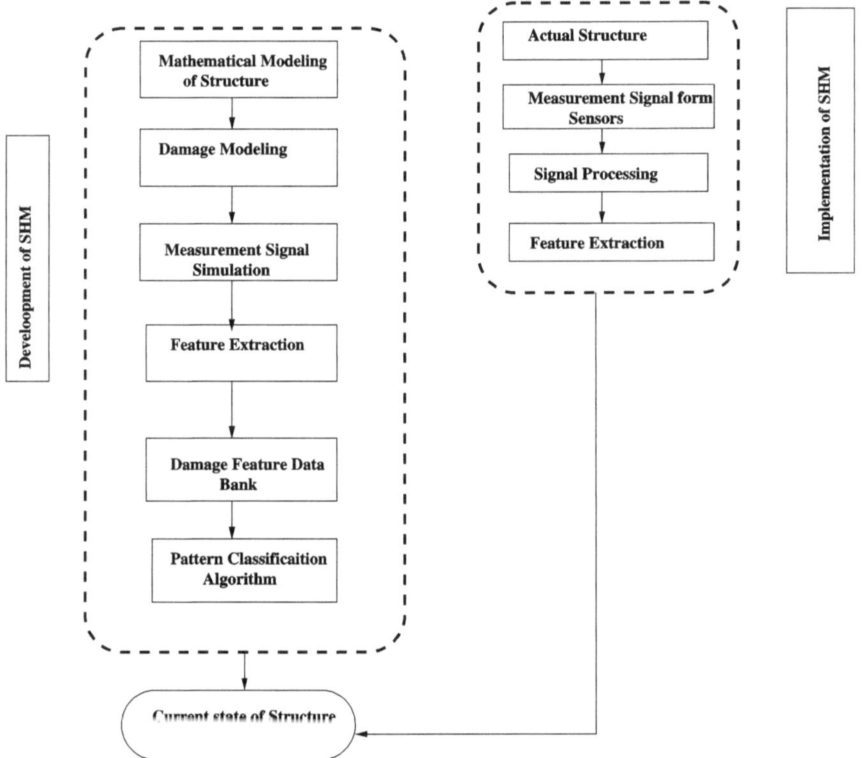

Fig. 1.1 SHM flowchart

involves two parts: SHM development process and SHM implementation process. These two processes go hand in hand, as the SHM development process depends upon the system environment, type of damage in the system, and types of signals measured. Precise mathematical modeling of a system is now possible due to dramatic improvements in the computational capability of modern computers. Accurate mathematical modeling of the damaged system behavior is an important part of model-based SHM. Modeling often involves the solution of the governing differential equations for the system with appropriate initial and boundary conditions. The governing differential equations are typically obtained by applying the laws of physics to the undamaged and damaged system.

The measurements which are used to track damage play a very important part in SHM. The type of measurement signals and locations is decided based on the type of damage, type of structure, operating environment of the structure, and level of accuracy needed for damage prediction. Structural engineering models are generally developed using finite element methods. Sometimes, actual experimental results and those predicted by finite element-based models may show discrepancies. These discrepancies can be minimized by improving the theory and by minimizing the discretization errors (insufficient mesh refinement, simplification of geometry) and numerical errors occurring during the resolution of the equations of motion. For example, short stubby beams are better modeled using Timoshenko beam theory, while long, slender beams are well modeled using Euler–Bernoulli beam theory. Thus, the modeling of short beams like turbine blades can be improved by simply switching from Euler–Bernoulli beam theory to Timoshenko beam theory. The problem of discretization errors can be addressed by doing a convergence study. The number of finite elements used to model the structure can be progressively increased and the value of the simulated damage indicator (for example, frequency or mode shape, etc.) can be monitored. The finite element mesh should be selected such that the damage indicator has converged. However, when localized damage is present in the structure (such as a crack in a beam), further local meshing is required to reduce the discretization error. Thus, care should be taken to ensure that the finite element model has converged in terms of the damage indicator for every possible damage simulation case considered. Discrepancies in the simulation of damage indicators can also arise from the uncertainties of the values of some parameters of the finite element model which are used as inputs. A finite element update procedure can be used to improve the modeling. Typically, model updating involves the solution of an optimization problem to find the optimal mass, stiffness, and damping matrices which lead to minimal differences between experimental and computed responses. Since physical models play a very important part in model-based diagnostics, considerable care should be taken to ensure that the baseline undamaged model and the damaged model are good representations of the real structure. Wherever possible, verification with experiments or with other modeling results is invaluable.

Material modeling is another important aspect of model-based SHM. Materials may not show the same behavior as that predicted by a theoretical model due to the nucleation and evolution of material damage. Material damage creates unacceptable and irreversible changes under various loading conditions or operating environments. Damage in materials leads to a reduction in some of the measurable properties such as strength, stiffness, or toughness. Damage modeling is a formidable task, as the relationships between changes in material response and changes in external factors, such as loading conditions or structural geometry, are generally complex. Material modeling depends on the boundary conditions or service environments imposed on the structure or material. Generally speaking, the more technologically advanced the application, the harsher the service environment and the more demanding the boundary conditions. Next, the type of material being modeled, traditional or nontraditional, single phase or multiphase, is an important factor in determining which model to use. Traditional materials are likely to have a well-established data base and associated material models, whereas nontraditional materials typically do not. For example, a localized damage in an isotropic material can be well modeled by a localized stiffness reduction using the continuum mechanics approach [9]. However, for a composite material, more elaborate models are needed, as the physics of damage evolution can be quite complex [10]. A third issue in damage modeling is that each type of material has a characteristic microstructure. Microstructure plays an important role in determining material behavior since the morphology of initial damage and the nucleation and evolution of load-induced damage are controlled by material microstructure. The fourth basic factor is the identification of the dominant forms of damage and the associated length scales at which they operate. Fifth, the evolution of damage needs to be considered, i.e., how it transitions through length scales and how it interacts with other forms of damage. As research in material modeling progresses, a better physical understanding of the underlying microstructure and damage evolution becomes possible. However, at any given time, SHM should use material and damage models which fit into the modeling framework. For example, models of cracks in materials should fit into a computationally efficient finite element framework. Models of damage in composites are easy to use if they can be expressed as changes to the laminated plate theory models. The SHM system designer has to make a tradeoff between physical detail and accuracy of the material model and its ease of incorporation into the finite element model.

The damage models can be classified as physics-based damage models and phenomenological damage models. The presence of flaws and defects in a structure creates a significant impact on material strength and toughness. Physics-based models are developed for predicting material behavior in the presence of flaws and cracks. These models make use of the stress field generated by a single crack or for a few idealized configurations of multiple cracks in a structure. Physics-based damage models involve continuum mechanics-based or fracture mechanics-based approaches. For example, a physics-based model can be an expression which gives the localized stiffness in a structure for a given crack size. Phenomenological failure theories attempt to predict when a material will fail based on some simple macroscopic criteria. In general, phenomenological failure theories do not generate both constitutive predictions and failure predictions and are not rate or history dependent. Generally speaking, phenomenological models are curve fits which are obtained from experimental data. An example of a phenomenological model is fatigue curves which link the material degradation to the number of cycles experienced by the structure. Typically, it is better to use physics-based models if they are available for a given damage and structure. However, in some cases, the damage mechanics can be very complex and recourse to a phenomenological model may be the only option.

The SHM approach can be best understood by looking at examples of real systems. One such system which is prone to failure is the helicopter. Unlike many other structures for which failure is only a cost issue, helicopter failure often results in human deaths. Therefore, developing an SHM capability for helicopters is very important; we will use this system for a case study in Chap 5 and introduce the helicopter rotor system problem here. The first step for developing an SHM system for a helicopter is to develop a mathematical model of the main rotor blade [11]. A comprehensive aeroelastic analysis based on finite elements in space and time is used to simulate a damaged helicopter rotor. Numerical results are obtained in hover and forward flight for the rotor. Selected predictions of rotor component loads are validated with flight test data. Fault models include moisture absorption, loss of trim mass, damaged pitch control system, defective lag damper, damaged trim tab, structural damage, and misadjusted pitch link. In this problem, relatively simple physics-based models are able to simulate the damage. Moisture absorption is modeled by increasing the mass of the blade. The loss of trim mass is modeled by a reduction in a mass at a small damage element. The misadjusted pitch link is modeled by a reduction in the rigid pitch angle of the damaged blade. The damaged pitch link control system is modeled by a reduction in the pitch link stiffness of the damaged blade. The defective lag damper is modeled by setting the damping coefficient of the lag damper of the damaged blade equal to zero. The damaged flap is modeled by a static shift in the deflection of the flap of the damaged blade. Further, the changes in the features of the rotor blades due to various damages are used to develop a damage prediction system using neural networks.

SHM problems involve the design of an algorithm which can detect and isolate damage from measurement data. The mathematical model is developed to simulate and study prospective measurements which can be used for monitoring the structural health. These prospective measurements or their signal processed versions then become the damage indicators. Since actual measurements are noisy, a pattern recognition algorithm needs to be developed to map the noisy measured data to the damage. For helicopter rotor blade health monitoring problems, a neural network can detect and quantify both single and multiple faults on the blade from noisy simulated data. Fuzzy logic is also used for damage detection in the helicopter rotor system [12]. In this fuzzy system, the measurement deviations due to damage are fuzzified and mapped to a set of faults using a fuzzy logic system. The output faults of the fuzzy logic system are four levels of damage (undamaged, slight, moderate, and severe) at five locations along the blade (root, inboard, center, outboard, tip). Even though this system is based on numerically simulated measurements, these measurements can be obtained experimentally [6, 13, 14] with reasonable accuracy, hence implementation of SHM becomes easy. A detailed discussion of neural networks and fuzzy logic is given later in this chapter and in the next chapter.

The rotor problem example illustrates some general ideas behind SHM. The model-based methods for SHM are developed by solving an inverse problem using various pattern recognition methods. The system response can be represented by an N-dimensional vector of real numbers, which represents a single point in an N-dimensional space. The simulated system response for different faults tends to be segregated into distinct regions of this space and can thus be regarded as patterns. Pattern recognition involves learning these patterns so that a given system response can be classified with a particular fault. Model-based damage detection methods are quite useful for accurate damage prediction using the inverse method, even with noisy data. The accuracy of the damage prediction is based on accurate damage modeling, accurate system modeling, and use of a pattern recognition approach. In this book, the advanced soft computing architecture of the genetic fuzzy system is presented as the pattern recognition tool for many SHM applications. Some of the applications involve the use of modal data for SHM. Therefore, modal-based methods are briefly discussed next.

1.2.2 Modal-Based Methods

Modal-based methods, which use frequency and mode shapes, are also widely used for damage detection [15]. Damage detection methods can be categorized based on the type of dynamic characteristics used and/or the techniques used to identify the damage from measured data. These dynamic characteristics include natural frequencies [16, 17], mode shapes [18, 19], mode shape curvatures [20, 21], and the frequency response function [22]. It is impossible to identify the locations and severity of damage in structures such as cantilever beams and helicopter rotor blades using frequency response function (FRF) data [23]. Traditional methods detect the location and severity of structural damage by minimizing the difference between test and analytic FRFs, which is a type of model updating or optimization method. Another approach is a time domain method that utilizes response signals in their purest form [24]. In this method, error magnification related to computational manipulations, e.g., an FFT, or similar procedures, is avoided. The method is applied to a clamped section of a helicopter rotor blade. Damage is induced by a mass increase (locally added weight) and a mass reduction (holes drilled into the structure), respectively. The experiments have shown that minimal effects on fundamental eigenfrequencies are reliably detected without any a priori knowledge of the structure.

For the modal-based damage identification methods, the most significant limitations are that they require accurate measurement of mode shapes and modal frequencies. These measurements are susceptible to sensor noise, contain errors inherent to curve fitting techniques, and require significant onboard processing capabilities. However, with the increasing accuracy of sensors and signal processing algorithms, the accuracy of frequency and mode shape measurements also continues to increase. Moreover, modal methods are not influenced much by small levels of damage, which is an advantage in preventing false alarms in damage-tolerant structures designed to take significant levels of damage. For instance, helicopter rotor blades can undergo significant damage prior to failure.

1.2.3 Localized Methods

Most of the model-based and modal-based damage detection methods are based on global measurements. Note that some modal-based methods are also model based, as they typically use a finite element model of the structure. However, some methods use only measurements and are model free [24]. Model-free methods are often good for finding the existence of damage, but have difficulties with damage isolation. For example, placing thresholds on sensor measurements is a model-free method. A few researchers proposed model-based and modal-based damage methods to give an approximate damage location. However, some faults, such as cracks and delamination, may not show up in global measurements unless they are large; therefore, localized measurements are necessary. Some of the prominent localized methods are discussed here with illustrative examples.

Acoustic emission is the most widely used localized damage detection method in aerospace applications. Acoustic emission is a passive approach in which an ultrasonic sensor is used to "listen" for the characteristic stress waves emitted as a crack propagates. For example, acoustic emission can be used for successful detection of cracks in rotor components such as non-rotating pitch links in a laboratory setting during fatigue tests [25].

Wave mechanics is another localized approach reported in the literature for damage detection in helicopter applications. This approach, in contrast to acoustic emission, is an active approach based on high frequency excitation. As ultrasound waves hit a discontinuity in the material, they are reflected. The wave mechanics approach, which is based on local continuum mechanics descriptions of the structural dynamics, uses analytically predicted scattering patterns to detect, locate, and characterize flaws. Such an approach can be used for detecting cracks in composite rotor flexbeams [26], and for detecting cracks and delamination in rotating flexbeams [27]. Flexbeams are torsionally soft beams attached near the blade hub and are used for twisting the rotor blades in bearingless helicopter rotors. SHM of such structures is very important for helicopter flight safety. Helicopter rotors have three key degrees of motion: flap bending, lag bending, and torsion. Early helicopters used hinges and bearings to alleviate the high stresses at the blade root. The advent of composite materials allowed the design of flexures which replaced the flap and lag hinge for the hingeless rotor. In the bearingless rotor, the pitch bearing is also replaced by an elastic flexbeam. However, a consequence of the use of composites in helicopter rotors is an increased propensity toward complex damage mechanisms. Metal structures have simpler modes of failure compared to composites. Therefore, the development of SHM systems for composites requires considerable work on the damage modeling aspects.

The localized techniques based on ultrasonic sensors may not work for thick-walled monolithic composites. One way to address this problem is to use computed tomography (CT) as a nondestructive test method for composite rotor blades such as those for the BO105, BK117, and EC135 in design and maintenance [28].

The electro-mechanical (E/M) impedance technique is another nondestructive evaluation (NDE) technology with high potential for in situ health monitoring of complex machinery. For example, the E/M impedance technique was used for detecting disbonds between adhesively bonded structural elements of helicopter blade sections from the Apache 64H helicopter [29]. Damage was mechanically induced in the structure in the form of local disbonds. An examination of the damaged-structure E/M impedance spectra and a comparison with the baseline spectra revealed some important phenomena such as frequency shift of existing peaks, increase in peak amplitudes, and the appearance of new peaks. It was found that the frequency shifts were consistently toward lower frequencies. This shift in frequency was explained by the increase in local compliance due to disbonds. The increased impedance amplitude was correlated with the decrease in local damping that appears when the two faying surfaces were separated. The appearance of new peaks was justified by the new local modes that were created when disbonds appear. However, this method requires further examination of the complex interaction between wave propagation, drive point impedance, structural damage, and electro-mechanical impedance of the piezoelectric wafer transducer.

The advent of smart materials such as piezoceramics has opened up many possibilities for SHM. Piezoceramic sensors and actuators and different techniques for damage detection have been used for health monitoring of a helicopter flexbeam [30]. Four different methods were used for testing the flexbeam: (a) the resonant comparison method using inertial actuator/PZT patches for longitudinal and transverse vibration, (b) the wave propagation method using inertial actuator/PZT patches for longitudinal and transverse vibration, (c) the acoustic emission (lead break) method using PZT sensor patches, and (d) the vibration response method using a PZT patch for bending excitation. The damage modeled in the flexbeam was delamination. It was found that the damage detection using longitudinal vibration was promising because the amplitude change of symmetric PZTs indicates the damage with minimal pre-damage data. Delamination that was perpendicular to the direction of wave propagation was possibly easier to detect using flexbeam vibration. The resonant comparison method using longitudinal vibration successfully detected delamination damage in a fiberglass beam. The method uses structural symmetry, no structural model is needed, the excitation does not need to be measured, and the diagnostic procedure is very repeatable and simple. However, this method needs to be validated for nonsymmetric structures.

The local damage detection methods are useful for accurate local damage prediction. Approaches such as those based on Lamb waves are especially attractive for aircraft composite structures [31]. However, the use of these methods for scanning damage in the whole structure is very time consuming and is not economical. A better strategy is to locate the damage to within a small area using a global damage detection method and then use the local methods to zoom in on the damage location.

1.3 Soft Computing Methods for Health Monitoring

SHM systems are typically developed by solving an inverse problem using the changes in some measurable properties of the structure to detect damage. Typically, models of the structure are used to simulate the damaged condition of the structure. The inverse problem becomes complicated because of the incomplete information and uncertainty in the modeling, measurements, and signal processing. There are several approaches for handling uncertainty in the inverse problems. The more traditional approach is to use estimation methods such as those based on Kalman filtering [32] or probabilistic methods such as those based on Bayesian reasoning [33, 34]. Another powerful approach which has recently gained popularity is the use of soft computing methods [35]. Therefore, the inverse problem for the development of an SHM system can be solved using soft computing methods like neural networks, genetic algorithms, and fuzzy logic to extract precise conclusions from slightly imprecise data. Soft computing differs from traditional hard computing in several important ways. Hard computing tries to find the exact answer, while soft computing tries to find an approximate answer. There are many problems in mathematics where hard computing comes immediately to mind, for example, in numerical analysis and mathematical logic. On the other hand, most decision making and evolutionary processes are soft computing processes and lead to good outcomes without being exact. For example, biological evolution proceeds randomly while being directed by the objective of survival of the fittest. However, it is clear from evolution that there is improvement in the species with the passage of time. The human brain also functions using a huge number of interconnected neurons and is able to perform complex pattern recognition tasks. Human beings are also able to arrive at many important decisions in their lives by using various fuzzy concepts such as "tall," "rich," "dangerous," etc. The pioneers of soft computing realized that new computing methods which address uncertainty and ambiguity are needed to address many real world problems. During the last two decades, the field of soft computing has grown considerably, and there are many papers and some journals which are devoted to this field. Some recent approaches of these soft computing methods for SHM are discussed next.

1.3.1 Neural Networks

The neural network is a powerful pattern recognition tool; it forms a nonlinear mapping between input and output data sets using training data. A network is developed using interconnected neurons. The inputs of each neuron consist of the weighted outputs of other neurons. A commonly used neural network is the feedforward neural network shown in Fig. 1.2. There is one input layer with n inputs, a hidden layer with m neurons, and one output layer with two outputs. For SHM problems, the two outputs typically represent damage size and location. The inputs are the damage indicators for the problem.

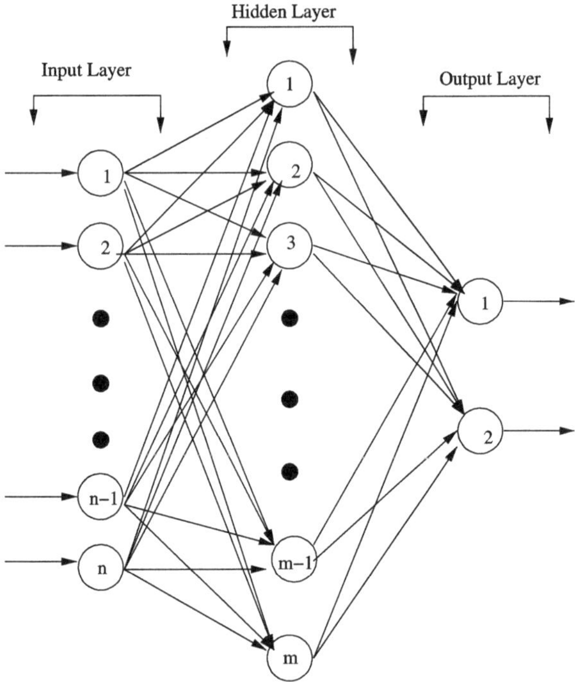

Fig. 1.2 Schematic representation of feedforward neural network

There can be one or more hidden layers between the input and output layers. Figure 1.2 shows a neural network with one hidden layer. In the feedforward neural network, data flows forward and each neuron receives inputs from the previous layer only. The neural network in Fig. 1.2 is thus a map from R^n to R^2 or the space of n real numbers to the space of 2 real numbers. The number of neurons in the input and output layer are fixed by the input and output vector of the pattern recognition problem. The number of neurons in the hidden layer plays a very important role in the function approximation capability of the neural network. An appropriate "optimal" selection of the number of neurons in the hidden layer can allow a neural network to approximate any given function with any desired level of accuracy.

The interconnections between the neurons have a weight associated with them which stores the functional mapping information in a distributed sense. The solution of the weights connecting the neurons is done through the process of learning by input-output data linking the vectors **x** and **y**. Once the neural network is trained, it develops the capability to generalize to points which were not included in the training set. The process of training the neural network is accomplished by using an optimization method which minimizes the error given by the neural network while using the weights as the design variables. A separate set of test data, which is known to be different from the training data, is used to test if the neural network can generalize well.

Generally, one hidden layer is sufficient for a feedforward neural network. Too few neurons in the hidden layer will make function approximation difficult. Too many neurons in the hidden layer will lead to overlearning, where the neural network loses its generalization capability. Selecting the appropriate number of neurons in the hidden layer is an art and is typically done using a trial-and-error procedure to ensure that the neural network performs well with training and testing data. Some input-output data should always be set aside for testing and should not be used for training. Test data is invaluable in evaluating the generalization capability of neural networks. If the network gives a low error with training data but does not perform well with test data, then overfitting is the likely cause.

The backpropagation algorithm is widely used for training feedforward neural networks. This method is based on gradient search. Associated with each neuron is an activation function. For example, the sigmoid function is widely used in practice for the neurons in the hidden layer,

$$f(x) = \frac{1}{1+e^{-x}}.$$

The above discussion is a very brief summary of neural networks; the subject is addressed in several textbooks [36–38]. The important thing to remember is that neural networks are multi-dimensional curve fits relating the output and input data.

Some recent examples of the use of neural networks for various applications related to the development of damage detection systems are as follows. A few researchers have developed damage detection systems for composite materials using neural networks. Mahapatra et al. [39, 40] developed a method for the estimation of damage configuration in a composite structure using acoustic wave propagation signal and a reduction-prediction neural network to deal with high-dimensional spectral data. A reduction-prediction network, which was a combination of an independent component analysis (ICA) and a multi-layer perceptron (MLP) neural network, was proposed to quantify the damage state related to transverse matrix cracking in composite laminates using an acoustic wave propagation model and the Fourier spectral response of the damaged structure under frequency band selective excitation. Further, Chakraborty [41] developed an approach for predicting the presence of embedded delaminations (in terms of their size, shape, and location) in fiber reinforced plastic composite laminates using natural frequencies as indicative parameters and an artificial neural network as a learning tool. The neural network was trained until the network learned to an acceptable level of accuracy using the ten modal frequencies obtained for various combinations of size, shape, and location of embedded delamination in a graphite/epoxy plate using a three-dimensional (3D) finite element model. It was found that the neural network can learn effectively about the size, shape, and location of the embedded delamination present in the laminate and can predict reasonably well when tested with an unknown data set.

Some researchers have used neural networks for prediction of fatigue damage and crack propagation. A backpropagation neural network for fatigue life prediction under multiaxial random loading was developed by Kang et al. [42]. The proposed neural network model was applied to the fatigue life prediction of an automotive subframe, which was subjected to 70 random load histories. The fatigue life data of the automotive subframe was obtained using a finite element model. The performance of the neural network model was evaluated by comparing outputs of the neural network with results of the conventional calculation method and was found to be acceptable in most fatigue design considerations. Al-Assaf and El-Kadi [43] demonstrated the use of neural networks for predicting the fatigue behavior of unidirectional glass fiber/epoxy composite laminae under tension-tension and tension-compression loading.

Online damage detection is another important area where neural networks have been used. Oberholster and Heyns [44] proposed a methodology for monitoring the online condition of axial flow fan blades with the use of neural networks. Results from a stationary experimental modal analysis of the structure were used for identifying global bladc mode shapes and their corresponding frequencies. It was demonstrated that it was possible to classify damage for several fan blades by using neural networks with online vibration measurements from sensors not necessarily installed on the damaged blades themselves. Yuan et al. [45] developed an online damage detection method applied to thin-walled composite structures adopting wideband Lamb waves based on active monitoring technology using a neural network. A new damage signature based on the acoustic-ultrasonic active monitoring technology was introduced to determine the presence and extent of delamination and impact damage in composites, while eliminating the influence of different distances between the actuator and sensor. A neural network was researched to take advantage of this new damage signature combined with several other signatures to decide the damage mode.

The above-mentioned references are a small sample of the literature on the use of neural networks in SHM. The literature on neural network-based damage detection systems shows that these methods have been demonstrated for various important problems such as composite structures, fatigue damage prediction, and online damage prediction. However, it is difficult to develop neural network-based damage detection for some of the critical cases where the system has insufficient training data and only human experience-based knowledge is available. Neural networks also require a huge amount of computer time for training, and a slight tuning of the neural network is difficult when there are few modifications in the model. One reason for the popularity of neural networks in SHM is the familiarity of structural engineers with the neural network method and the availability of software packages. However, there is considerable literature which shows that damage detection is a problem plagued with uncertainty and ambiguity and that linguistic outputs in terms of damage location and size are preferable for maintenance actions. In this book, we shall see that the fuzzy logic method provides a superior alternative to the neural network approach. However, fuzzy logic requires an element of learning to give accurate results. This learning function can be framed as an optimization problem and solved using genetic algorithms, which are briefly discussed in the next section.

1.3.2 Genetic Algorithms

The genetic algorithm (GA) is another popular soft computing tool used for damage detection. GAs are search algorithms developed using the principles inspired by natural selection to evolve solutions to optimization problems. Most of the GA-based damage detection methods are developed for minimizing an error measure based on the damage indicator values of the undamaged structure. The design variables are indicative of the state of structural health such as the stiffness or mass distributions. The GA approach to damage detection is typical of system identification methods which were used in the early years of damage detection research and sought to find the damaged structure which matches the measurements. The GA-based methods differ from each other based on the objective function formation.

GAs are optimization algorithms which maximize a fitness function. They are based on the evolutionary concept of Darwin that populations become fitter through a process of natural selection and survival of the fittest. A fundamental difference between GA methods and traditional gradient-based methods is that the GA starts from a population of points, while traditional methods start from one point. More importantly, the GA does not use gradient information to move in the design space and is therefore also called a derivative-free or zero-order method. The calculation of derivatives can be difficult for some functions. Also, gradient-based algorithms have numerical problems in regions of the design space where the Hessian matrix is ill conditioned or not positive definite. The Hessian matrix is composed of the second derivatives of a multivariate function. For an SHM problem, it is difficult to know a priori how the derivatives of the objective function will behave in different regions of the design space. Another advantage of GAs is their ability to find a global minimum point. In some design spaces, there is more than one minimum point where the gradient of the function becomes zero and the second derivative or the Hessian is positive. Gradient-based methods tend to get stuck at a nearby local minimum point. In complex problems, it is very difficult, if not impossible, to know where the global minimum may lie in the design space without doing considerable mathematical and computational analysis. A particular type of function which guarantees that there is a global minima is the convex function, which has a positive definite Hessian matrix. However, the complicated numerically evaluated functions which occur in SHM are rarely convex, and it is very difficult to prove convexity. Therefore, a GA is often used in damage detection applications, as it offers reduced complexity in terms of mathematical requirements on gradients and their derivatives.

The most common GA is the binary GA, where the design variables are represented as binary strings (such as 1001001). Operations of mating pool selection, crossover, and mutation are performed on the binary strings to simulate the reproduction process. Note that evolution involves many successful reproduction events. Strings which are fitter in terms of the fitness function are forwarded for propagation to the next generations, while unfit strings are removed. The creation of the starting population, selection of the mating pool, and the crossover and mutation operations make use of random numbers. Therefore, GAs are also called probabilistic or stochastic optimization methods. Any two runs of a GA code are different because of the presence of random numbers in the algorithm.

Some examples of the use of GAs for damage detection are discussed here. Rao et al. [46] proposed a method for locating and quantifying the damage in structural members using the concept of residual forces. To describe the damage in a structure, finite element models were parameterized by structural stiffness reduction parameters. The damage parameters were determined by minimizing a global error derived from dynamic residual vectors, which were obtained by introducing simulated "experimental" data into the eigenproblem. An eigenvalue prediction algorithm, along with a normalized residual function, was employed to formulate the objective function.

Nag et al. [47] proposed an efficient strategy for identification of delamination in composite beams and connected structures using a GA integrated with finite element code for automation. A spectral finite element model consisting of a damaged spectral element was used for model-based prediction of the damaged structural response in the frequency domain. Another approach based on Lamb waves and a projection GA for damage detection in composite plates was proposed by Xu et al. [48]. This method first formulates the damage detection problem as an optimization problem of minimizing the error between the measured and calculated surface displacement response derived from Lamb waves. Numerical examples were presented to verify the proposed method for detection of cracks inside composite plates.

Most of the GA-based SHM methods are based on the error reduction between the measurements obtained using an analytical or numerical model and the experimental results. A few researchers also used modified GA versions such as micro GA (muGA) and projection GA. A disadvantage of this method is that GAs are very time consuming and therefore difficult to use in an online setting; they are mostly intended for offline use. Another problem is they do not directly account for noisy data. Some recent applications of GA to SHM advocate a damage penalization function to address the issue of noisy data [49]. In this approach, the objective function not only searches for the best correlation but also for the minimum possible damage. Thus, it is possible to use a GA to address noisy data by modifying the objective function of adding constraints. However, fuzzy logic may offer a conceptually simpler and more elegant approach to handling uncertainty in SHM problems.

1.3.3 Fuzzy Logic System

A fuzzy system is any fuzzy logic-based system which uses fuzzy logic as the basis for the representation of the different forms of knowledge. Fuzzy systems may also use fuzzy logic to model the interactions and relationships among the system variables. Fuzzy systems address uncertainty directly by using linguistic reasoning, which is more robust to uncertainty than pure numerical reasoning. Thus, fuzzy logic converts a number such as the height of a person (say 6 ft) to a word (tall). While a height of 6 ft is an exact number, the concept "tall" is fuzzy and ambiguous in the sense that many people with a range of heights can be called "tall." Fuzzy systems try to work with concepts expressed in words. This number-to-word transformation is a key step which gives fuzzy systems the power to handle uncertainty.

Recent examples of the use of fuzzy logic systems for developing damage detection systems are discussed next. Since fuzzy logic forms a key component of this book, an introduction to this subject is provided in the next chapter in some detail. A brief discussion of some literature on the application of fuzzy logic to SHM is given next.

Sawyer and Rao [50] presented a procedure for structural fault detection based on fuzzy logic. Fuzzy logic and continuum damage mechanics were used to process and analyze the uncertainties and complexities of damaged structures. Damage in the structure was modeled by a localized reduction in stiffness. Fuzzy associations between observable structural responses and damage conditions were generated by finite element simulations. The fuzzy associations or rules were encoded in a fuzzy associative memory bank to form a knowledge base. This knowledge base was referenced by a fuzzy inference algorithm which infers possible damage locations and levels based on the evidence provided by changes in the structural states. The numerical examples demonstrated the performance advantages of the fuzzy logic-based system in noisy or uncertain conditions.

Dempsey and Afjeh [51] developed a diagnostic tool for detecting damage to spur gears using fuzzy logic with two different measurement technologies: wear debris analysis and vibration. This diagnostic tool was developed and evaluated experimentally by collecting vibration and oil debris data from fatigue tests performed in the NASA Glenn Spur Gear Fatigue Test Rig. A simple model was defined by the fuzzy rules and the membership functions for the experiments when pitting damage occurred. The ability to define valid ranges and limits for each membership function was found to be critical to the success of the model at predicting damage. Vibration data were collected from accelerometers and used in previously validated gear vibration diagnostic algorithms. Oil debris data were collected using a commercially available oil debris sensor. Oil debris and vibration data were integrated using fuzzy logic analysis techniques to predict damage in the spur gear. From the results it was observed that the use of two measurement technologies together improved the detection of pitting damage on spur gears.

The fuzzy-based methods are also applied to civil engineering structures for extracting the data obtained from NDT techniques for developing an SHM system. Soh and Bhalla [52] proposed a fuzzy probability-based damage model based on the extracted equivalent stiffness to evaluate the extent of damage using impedance data. This enabled the calibration of the piezo-impedance transducers in terms of damage severity, thus serving as a practical empirical phenomenological damage model for quantitatively estimating damage severity in concrete. Zhao and Chen [53] presented a fuzzy rule-based inference system for bridge damage diagnosis and prediction, which aimed to provide bridge designers with valuable information about the impacts of design factors on bridge deterioration. A modified mountain clustering method was employed to create the training data set. In the mountain clustering method, each cluster center becomes a prototypical data point that exemplifies a characteristic behavior of the system. A fuzzy partitioning algorithm was implemented to construct the membership functions of the input variables and to deduce the fuzzy rules from the numerical data. The generated rule base was checked and optimized based on the similarity measures among the input fuzzy sets. The use of fuzzy logic in addressing uncertainty issues in SHM is increasing [54]. In general, uncertainties can cause false diagnosis and imprecise prognosis if not taken into account, and therefore a soft computing method must be used for pattern recognition.

Two major advantages of a fuzzy logic-based damage detection system are that the system can be developed even with a lower number of data sets and that the system can be tuned with the upgraded model. Another advantage of fuzzy logic is that the human experience-based knowledge can be utilized to develop or tune the system when sufficient data is not available. However, the tuning of the fuzzy system is a manual process and is therefore time consuming and not optimal. As fuzzy logic systems do not have the capability of learning from data, it is difficult to develop a fuzzy system when a large number of discrete data is available as a knowledge base. In contrast to the large amount of work available on SHM using neural networks, much less work has been done on the use of fuzzy logic in SHM. This is largely due to the unfamiliarity of structural engineers with fuzzy logic. This book aims to remove this gap in the training of structural engineers working on SHM systems.

1.3.4 Hybridized Soft Computing Methods

Each of the soft computing methods, neural networks, genetic algorithms, and fuzzy logic, offers its own advantages and suffers from certain weaknesses. Although all three of these methods share some common characteristics, they are considered complementary, as desirable features lacking in one approach are present in another. There was an initial phase of research in which the soft computing methods were applied in isolation. This was followed by a second phase of research on hybrid systems obtained by combining these soft computing methods. The advantages of various combinations of these soft computing methods are given below [55, 56].

Fuzzy Logic and Neural Networks: The idea of combining these two soft computing methods originates from two observations. (1) Fuzzy logic systems are not capable of learning, adaptation, or parallel computation. On the other hand, these characteristics are clearly present in neural networks. (2) Neural networks lack flexibility, human interaction, or knowledge representation. However, these characteristics lie at the core of fuzzy logic systems. Depending on which component dominates in a hybrid system, one can distinguish between neuro-fuzzy systems and fuzzy-neural network systems. Typically, neuro-fuzzy systems have emerged as a popular choice for SHM problems.

Neural Networks and Genetic Algorithms: This combination is typically used for two main applications: (1) a GA is used to optimize the parameters of the learning method that adapts the synaptic weights of the neural network, and (2) assuming a bounded number of hidden neurons, GAs are applied to find the network structure that minimizes the number of training cycles. Since neural networks are often trained using the backpropagation algorithm, which is an optimization procedure which relies on gradients, the use of a GA in the evaluation of the weights of the neural network and the architecture is feasible. However, the use of a GA as an optimization tool for neural network development can be computationally intensive and is not well suited to SHM problems.

Genetic Algorithms and Fuzzy Logic: GAs provide robust search capabilities allowing global and local search in complex design spaces. A GA is a derivative-free optimization method which can find a global maximum for a fitness function while avoiding problems related to function continuity and differentiability. Fuzzy systems present robust and flexible methods in problems which are subject to imprecision and uncertainty. The linguistic representation of knowledge permits a person to interact with the fuzzy system in a straightforward manner. In fact, fuzzy logic tries to use the linguistic logic of computing with words which humans use to arrive at decisions. This "fuzzy" logic is different from the black-and-white classical logic which is used in philosophy and mathematics. The hybridization of fuzzy logic and GAs gives rise to two methods: fuzzy evolutionary algorithms and genetic fuzzy systems. Fuzzy evolutionary algorithms are GAs whose parameters such as fitness function and stopping criteria are fuzzified. Such systems take advantage of a tolerance for imprecision in order to save computational resources. On the other hand, in a genetic fuzzy system, a GA evolves a fuzzy logic system by tuning the variables describing fuzzy membership functions and by learning fuzzy rules. The genetic fuzzy system can automate the process of developing fuzzy systems by treating various parameters of the fuzzy system as design variables which are obtained from the optimization process by the GA. If the optimization process is set up to maximize the success rate for damage detection, the genetic fuzzy system can result in an optimal SHM system. It thus avoids the problems of neural networks and fuzzy systems which tend to be ad hoc in nature and whose creation is more of an art depending on the skill of the designer than a science. We will see in this book that the genetic fuzzy system represents a very powerful tool for SHM.

Very few researchers have used hybridized soft computing methods for solving damage detection problems. Ramu and Johnson [57] developed a novel method by integrating the concepts of fuzzy logic with neural networks for damage detection applications. The main feature of this integrated approach using fuzzy logic and neural networks was that it requires no additional effort at the training level, but it can be used against fuzzy input at the end user level with the integration of the vertex method. The vertex method was a concept for discretization of membership value domains of variables. In addition, a modular development of the network was proposed which made the training fast and efficient. The above approach was demonstrated using a composite plate modeled using the finite element method. The damage occurring due to absorbed moisture and elevated temperature was modeled as a reduction in material properties and loss of support fixity. It was found that the hybridized neural network approach was a very promising tool for the damage assessment task.

Wang [58] developed a new neuro-fuzzy diagnostic system for gear systems by integrating the strength of robust signal processing techniques. The test results demonstrate that the novel neuro-fuzzy system, because of its adaptability and robustness, significantly improves the diagnostic accuracy. It outperforms other related classifiers, such as those based on fuzzy logic and neuro-fuzzy schemes, which adopt different types of rule weights and employ different training algorithms.

Lee and Lam [59] presented the application of a novel neural network model for the diagnosis of structural damage. The neural network model, which was denoted the GRNNFA, was a hybrid model combining the General Regression Neural Network (GRNN) model and the Fuzzy ART (FA) model. It not only retained the important features of the GRNN and FA models (i.e., fast and stable network training and incremental growth of network structure), it also facilitated the removal of noise embedded in the training samples. The measured modal parameter changes due to a particular damage were treated as patterns for that damage. The proposed diagnosis method was demonstrated using a ten-story shear building model. The damage was modeled by inter-story stiffness reduction. The proposed GRNNFA model was trained to learn those patterns in order to detect the possible damage location of the structure. The results of this study demonstrated the feasibility of applying the GRNNFA model to structural damage diagnosis even when the training samples were noise contaminated.

Zio and Gola [60] applied the neuro-fuzzy approach to fault diagnosis in rotating machinery. Faults in such machines can lead to considerable downtime and losses and must be found promptly and accurately. An initial fuzzy knowledge base is obtained empirically with bearing vibration used as the damage indicator, and then the algorithm modifies the rule base for more accurate classification. It was found that the method provided good classification of faults while keeping the system interpretable.

The new trend of hybridizing the best features of various soft computing methods gives a tool for developing more accurate damage detection algorithms. The above combinations of various soft computing algorithms can be chosen based on the type of data available and the output requirements. In this book, one of the advanced hybridized soft computing algorithms, the genetic fuzzy system, is developed for damage detection and is demonstrated for SHM of beam-type structures and for a composite helicopter rotor blade.

1.4 Book Summary

This first chapter has given a background of SHM along with various SHM approaches and an overview of the relevant literature along with some recent review papers. In the second chapter, an introduction to genetic algorithms and fuzzy logic systems is given. Further, Chap. 2 explains the development of the genetic fuzzy system (GFS) by the hybridization of fuzzy logic and the genetic algorithm. In Chap. 3, the development of the SHM system is introduced with an example of an isotropic cantilever beam. The beam is an ubiquitous structural member, and its damage detection is a fundamental problem in SHM. This chapter serves as a tutorial introduction to the application of the GFS for SHM. Chapter 4 demonstrates the use of the GFS for SHM for a more complicated structure, the hollow composite tube, in which

SHM will predict the level of a damage known as matrix cracking. Such composite tubes are used for power transmission poles. Thus, Chap. 4 involves an increase in mathematical model complexity of the structure and the damage and shows the application of the GFS to a more realistic structure. Chapter 5 demonstrates use of the GFS for developing global and local SHM systems for a composite helicopter main rotor blade in forward flight. The helicopter is a complicated mechanical system which is susceptible to damage due to high levels of vibration and dynamic stresses. This chapter provides an application to a realistic aerospace system using aeroelastic analysis of damaged rotor blades as the mathematical model and complicated damage evolution mechanics covering matrix cracking, delamination, and fiber breakages. By considering the application of the GFS to three different structures with increasing levels of complexity, this book provides the reader with the knowledge needed to apply the GFS to his or her SHM problem.

References

1. Brian, D.: Larger helicopter HUM/FDR: benefits and developments. In: Proc. 55th Annual Forum of the American Helicopter Society, Montreal, Canada, pp. 1839–1846 (1999)
2. Cronkhite, J.: Practical application of health and usage monitoring (HUMS) to helicopter rotor, engine and drive system. In: Proc. 49th Annual Forum of the American Helicopter Society, St. Louis, MO, USA, pp. 1445–1455 (1993)
3. Land, J., Weitzman, C.: How HUMS have the potential of significantly reducing the direct operating costs for modern helicopters through monitoring. In: Proc. 51st Annual Forum of the American Helicopter Society, Fort Worth, TX, USA, pp. 744–757 (1995)
4. Pouradier, J., Trouvi, M.: An assessment of Eurocopter experience in HUMS development and support. In: Proc. 57th Annual Forum of the American Helicopter Society, Alexandria, VA, USA, pp. 1790–1797 (2001)
5. Stevens, P., Hall, D., Smith, E.: A multidisciplinary research approach to rotorcraft health and usage monitoring. In: Proc. 52nd Annual Forum of the American Helicopter Society, Washington, DC, USA, pp. 1732–1751 (1996)
6. Cleveland, G., Trammel, C.: Integrated health and usage monitoring system for the SH-60B helicopter. In: Proc. 52nd Annual Forum of the American Helicopter Society, Washington, DC, USA, pp. 1767–1787 (1996)
7. Kumar, S., Roy, N., Ganguli, R.: Monitoring low cycle fatigue damage in turbine blades using vibration characteristics. Mech. Syst. Signal Process. **21**, 480–501 (2007)
8. Roy, N., Ganguli, R.: Helicopter rotor blade frequency evolution with damage growth and signal processing. J. Sound Vib. **283**, 821–851 (2005)
9. Beena, P., Ganguli, R.: Structural damage detection using fuzzy cognitive maps and Hebbian learning. Appl. Soft Comput. **11**, 1014–1020 (2011)
10. Gayathri, P., Umesh, K., Ganguli, R.: Effect of matrix cracking and material uncertainty on composite plates. Reliab. Eng. Syst. Saf. **95**, 716–728 (2010)
11. Ganguli, R., Chopra, I., Haas, D.: Detection of helicopter rotor system simulated faults using neural networks. J. Am. Helicopter Soc. **42**, 161–171 (1997)
12. Ganguli, R.: Health monitoring of helicopter rotor in forward flight using fuzzy logic. AIAA J. **40**, 2773–2781 (2002)
13. Miles, T., Lucas, M., Halliwell, N., Rothberg, S., et al.: Torsional and bending vibration measurements on rotors using laser technology. J. Sound Vib. **266**, 441–467 (1999)
14. Cabell, R., Fuller, C., O'Brien, W.: Neural network modelling of oscillatory loads and fatigue damage estimation of helicopter components. J. Sound Vib. **209**, 329–342 (1998)

15. Doebling, S., Farrar, C., Prime, M.: A summary review of vibration based damage identification methods. Shock Vib. Dig. **30**, 91–105 (1998)
16. Salawu, O.: Detection of structural damage through changes in frequency: a review. Eng. Struct. **19**, 718–723 (1997)
17. Diazvaldes, S., Soutis, C.: Delamination detection in composite laminates from variations of their modal characteristics. J. Sound Vib. **228**, 1–9 (1999)
18. Leonard, F., Lanteigne, J., Lalonde, S., Turcotte, Y., et al.: Free vibration behavior of a cracked beam and crack detection. Syst. Signal Process. **15**, 529–48 (2001)
19. Zou, Y., Tong, L., Steven, G.: Vibration based model dependant damage (delamination) identification and health monitoring for composite structures: a review. J. Sound Vib. **230**, 357–78 (2000)
20. Wahab, A.: Effect of modal curvatures on damage detection using model updating. Mech. Syst. Signal Process. **15**, 439–45 (2001)
21. Ratcliffe, C.: A frequency and curvature based experimental method for detecting damage in structures. J. Vib. Acoust. **122**, 32–49 (2000)
22. Sampaio, R., Maia, N., Silva, J.: Damage detection using frequency response function curvature method. J. Sound Vib. **226**, 1029–42 (1999)
23. Hwang, K.C.: Damage detection in structures using a few frequency response measurements. J. Sound Vib. **2701**, 1–14 (2004)
24. Cattarius, J., Inman, D.: Experimental verification of intelligent fault detection in rotor blades. Int. J. Syst. Sci. **31**, 1375–1379 (2000)
25. Schoess, J., Malver, F., Iyer, B., Kooyman, J., et al.: Rotor acoustic monitoring system (RAMS): a fatigue crack detection system. In: Proc. 53rd Annual Forum of the American Helicopter Society, Virginia Beach, VA, USA, pp. 274–281 (1997)
26. Lakshmanan, K., Pines, D.: Damage identification of chordwise crack size and location in uncoupled composite rotorcraft flexbeams. J. Intell. Mater. Syst. Struct. **9**(2), 146–155 (1998)
27. Purekar, A., Lakshmanan, K.: Detecting chordwise cracks and delamination in uncoupled composite rotorcraft flexbeams under rotation. In: Proc. 54th Annual Forum of the American Helicopter Society, Washington, DC, USA, pp. 1026–1043 (1998)
28. Oster, R.: Computed tomography as a non-destructive test method for fiber main rotor blades in development, series and maintenance. In: Computerized Tomography for Industrial Applications and Image Processing in Radiology, Berlin, Germany (1999). (Paper 4)
29. Giurgiutiu, V., Rogers, C.: Recent advancements in the electro-mechanical (E/M) impedance method for structural health monitoring and NDE. In: The SPIE's 5th Annual International Symposium on Smart Structures and Materials, Catamaran Resort Hotel, CA (1998). (Paper 3329-53)
30. Ghoshal, A., Harrison, J., Sundaresan, M., Hughes, D., Schulz, M., et al.: Damage detection testing on a helicopter flexbeam. J. Intell. Mater. Syst. Struct. **12**, 315–330 (2001)
31. Diamanti, K., Soutis, C.: Structural health monitoring techniques for aircraft structures. Prog. Aerosp. Sci. **46**, 342–352 (2010)
32. Alkahe, J., Oshman, Y., Rand, O.: Adaptive estimation methodology for helicopter blade structural damage detection. J. Guid. Control Dyn. **25**, 1049–1057 (2002)
33. Gelsema, E.: Diagnostic reasoning based on a genetic algorithm operating in a Bayesian belief network. Pattern Recognit. Lett. **17**, 1047–1055 (1996)
34. Lucas, P.: Bayesian model-based diagnosis. Int. J. Approx. Reason. **27**, 99–119 (2001)
35. Worden, K., Staszewski, W.J., Hensman, J.J.: Natural computing for mechanical systems research: a tutorial review. Mech. Syst. Signal Process. **25**, 4–111 (2011)
36. Simon, H.: Neural Networks and Learning Machines, 3rd edn. Pearson Education, Upper Saddle River (2009)
37. Kevin, G.: An Introduction to Neural Networks. Taylor and Francis, London (2003)
38. Laurene, V.F.: Fundamentals of Neural Networks: Architectures, Algorithms, and Applications. Prentice-Hall, Englewood Cliffs (1994)
39. Mahapatra, D., Suresh, S., Omkar, S., Gopalakrishnan, S., et al.: Estimation of degraded laminate composite properties using acoustic wave propagation model and a reduction–prediction network. Eng. Comput. **22**, 849–876 (2005)

40. Garg, A.K., Mahapatra, D., Suresh, S., Gopalakrishnan, S., Omkar, S.: Estimation of composite damage model parameters using spectral finite element and neural network. Compos. Sci. Technol. **64**, 2477–2493 (2004)
41. Chakraborty, D.: Artificial neural network based delamination prediction in laminated composites. Mater. Des. **26**, 1–7 (2005)
42. Kang, J., Choi, B., Lee, H., Kim, J., Kim, K., et al.: Neural network application in fatigue damage analysis under multiaxial random loadings. Int. J. Fatigue **28**, 132–140 (2006)
43. Al-Assaf, El-Kadi: Fatigue life prediction of unidirectional glass fiber/epoxy composite laminae using neural networks. Compos. Struct. **53**, 65–71 (2001)
44. Oberholster, A., Heyns, P.: On-line fan blade damage detection using neural networks. Mech. Syst. Signal Process. **20**, 78–93 (2006)
45. Yuan, S., Wang, L., Peng, G.: Neural network method based on a new damage signature for structural health monitoring. Thin-Walled Struct. **43**, 553–563 (2005)
46. Rao, M., Srinivas, J., Murthy, B.: Damage detection in vibrating bodies using genetic algorithms. Comput. Struct. **82**, 963–968 (2004)
47. Nag, A., Mahapatra, D., Gopalakrishnan, S.: Identification of delamination in composite beams using spectral estimation and a genetic algorithm. Smart Mater. Struct. **11**, 899–908 (2002)
48. Xu, Y., Liu, G., Wu, Z.: Damage detection for composite plates using lamb waves and projection genetic algorithm. AIAA J. **40**, 1860–1866 (2002)
49. Meruane, V., Heylen, W.: An hybrid real genetic algorithm to detect structural damage using modal properties. Mech. Syst. Signal Process. **25**, 1559–1573 (2011)
50. Sawyer, J., Rao, S.: Structural damage detection and identification using fuzzy logic. AIAA J. **38**, 2328–35 (2000)
51. Dempsey, P., Afjeh, A.: Integrated oil debris and vibration gear damage detection technologies using fuzzy logic. J. Am. Helicopter Soc. **49**, 109–116 (2004)
52. Soh, C., Bhalla, S.: Calibration of piezo-impedance transducers for strength prediction and damage assessment of concrete. Smart Mater. Struct. **14**, 671–684 (2005)
53. Zhao, Z., Chen, C.: A fuzzy system for concrete bridge damage diagnosis. Comput. Struct. **80**, 629–641 (2002)
54. Lopez, I., Sarigul-Klijn, N.: A review of uncertainty in flight vehicle structural damage, monitoring, diagnosis and control: challenges and opportunities. Prog. Aerosp. Sci. **46**, 247–273 (2010)
55. Cordon, O.: A historical review of evolutionary learning methods for Mamdani-type fuzzy rule based systems: designing interpretable genetic fuzzy systems. Int. J. Approx. Reason. **52**, 894–913 (2011)
56. Jiang, S.F., Zhang, C.M., Zhang, S.: Two-stage structural damage detection using fuzzy neural networks and data fusion. Expert Syst. Appl. **38**, 511–519 (2011)
57. Ramu, S., Johnson, V.: Damage assessment of composite structures using fuzzy logic integrated neural-network approach. Comput. Struct. **57**, 491–502 (1995)
58. Wang, W., Ismail, F., Golnaraghi, F.: Neuro-fuzzy approach to gear system monitoring. IEEE Trans. Fuzzy Syst. **12**, 710–723 (2004)
59. Lee, E., Lam, H.: NN-based structural damage diagnosis using measured vibration data. In: Proceedings of the Knowledge-Based Intelligent Information and Engineering Systems. Part 3, vol. 3215, pp. 373–379 (2004)
60. Zio, E., Gola, G.: A neuro-fuzzy technique for fault diagnosis and its application to rotating frequency. Reliab. Eng. Syst. Saf. **94**, 78–88 (2009)

Chapter 2
Genetic Fuzzy System

The damage detection problem is a pattern classification problem which is based on ambiguous, noisy, or missing input information. The input information is typically obtained from sensors placed on the structure or embedded in the structure. Sensor measurements are often subjected to some processing before being used as damage indicators. Though considerable effort is made to find damage indicators which show a high degree of sensitivity to the damage size and location, the problems of ambiguity, imprecision, and noise present in the measured data are likely to propagate into the damage indicators. In fact, in some cases, the signal processing performed on the sensor measurements may increase the noise level in the damage indicators. As an example, differentiation of data to get a damage indicator can amplify the noise contamination.

Based on this information from sensors, a damage detection system should provide definite outputs to help the maintenance engineers. For example, the following questions need to be answered.

1. Is there damage in the system?
2. Where is the damage?
3. What is the damage size?
4. What needs to be done?
5. How much longer can the structure be used?

The answers to these questions should be as accurate as possible. Answers expressed in words instead of in numbers are often more useful for maintenance engineers. Among the several soft computing methods, fuzzy logic is the one which maps numerical inputs into linguistic outputs. However, fuzzy logic lacks the capability of learning from the given data, and the rules which govern the fuzzy system must be developed by human experts. This process of developing the fuzzy rule base is difficult and can become very complicated if the number of inputs and outputs increases. The hybridization of fuzzy logic with a genetic algorithm gives an advanced soft computing algorithm called the genetic fuzzy system (GFS), which generates the fuzzy rules automatically from the data. In this chapter, the terms used in the formulation of a GFS for a damage detection problem are explained. The first step in developing the GFS is to understand the concepts underlying fuzzy logic.

P.M. Pawar, R. Ganguli, *Structural Health Monitoring Using Genetic Fuzzy Systems*,
DOI 10.1007/978-0-85729-907-9_2,

2.1 Fuzzy Logic System

Fuzzy logic is a unique soft computing method which simultaneously handles numerical data and linguistic knowledge. This unique feature has led it to be called "computing with words." People use fuzzy logic all the time to arrive at decisions in the complex settings in which they operate. For instance, we may conclude that a "fat" person who is "sedentary" is likely to have "health problems." In this case, the words "fat" and "sedentary" represent fuzzy concepts. These concepts could be numerically quantified using exact weight bounds and exercise bounds. However, human reasoning does not operate using such numbers and still often reaches surprisingly accurate conclusions using fuzzy rules. Another important fact in fuzzy logic is the degree of membership. For example, a person may have "fatness" between the levels of 0 and 1. He or she may also be "sedentary" between the levels of 0 and 1. A fuzzy rule could then be stated as: If a person is "fat" and "sedentary" then he or she has "health problems." We can see that an extremely overweight person who is completely sedentary is "very likely" to have health problems. A marginally "fat" person who is completely sedentary may also have some health problems. A very "fat" person who is not "sedentary" may also have some health problems. Finally, a much less "fat" person who is much less "sedentary" may have far fewer health problems. Thus, fuzzy logic can address a variety of situations using the concept of degree of membership in a fuzzy set. Thus the membership in the fuzzy sets "fat" and "sedentary" is not binary (1 and 0) but can have any number between 0 and 1. This simple example clearly shows how humans constantly assign fuzzy memberships to concepts to perform decision making. The output in this case clearly represents a diagnosis of health.

We see that a key feature of fuzzy logic is the use of words for computing. Other soft computing methods such as neural networks and genetic algorithms typically work with numerical data only. In other words, they convert a set of numbers to another set of numbers. However, fuzzy logic converts a set of words into another set of words. Expert systems can also work with words but use crisp logic in the sense that the concepts can only be either 1 or 0. So the rules of expert systems are similar to if-then-else rules in computer programs. Expert systems tend to deteriorate very rapidly in the presence of noise in the data. Despite these differences, fuzzy systems can be interpreted as generalizations of rule-based expert systems with the binary logic framework being replaced by fuzzy logic.

A fuzzy logic system is a nonlinear mapping of an input feature vector into a scalar output [1]. Fuzzy set theory and fuzzy logic provide the framework for the nonlinear mapping. Fuzzy logic systems have been widely used in engineering applications, because of the flexibility they offer designers and their ability to handle uncertainty. A fuzzy logic system can be expressed as a linear combination of fuzzy basis functions and is a universal function approximator. A schematic diagram of fuzzy logic is shown in Fig. 2.1 for $\Delta\omega$ as crisp inputs and *damage size* and *damage location* as crisp outputs. Here, $\Delta\omega$ is a measurement delta which can be a change in frequency between the undamaged and damaged structure, or a change in the strain output measured by a piezoelectric sensor, for example. The measurement

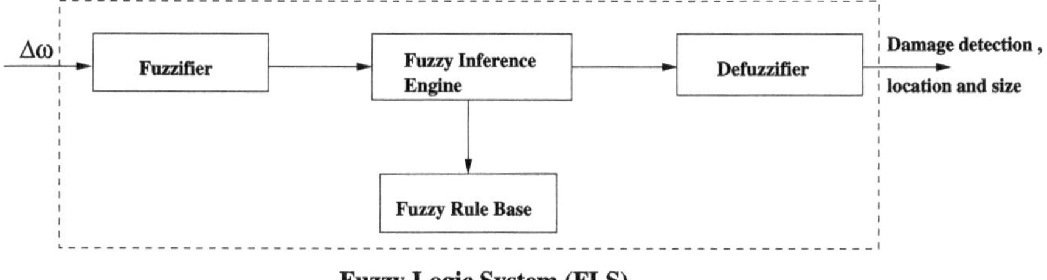

Fig. 2.1 Schematic representation of fuzzy logic system for SHM

deltas could also be changes in strains, blade tip responses, and blade root loads in a helicopter rotor blade. A very condensed introduction to fuzzy logic is provided below. Further information on fuzzy logic systems is available from textbooks [1–3]. Though fuzzy logic has become a vast topic, its application for structural health monitoring (SHM) requires the knowledge of only a few key concepts, which are outlined next.

A typical multi-input single-output (MISO) fuzzy logic system performs a mapping from a set $V \in R^m$ to $W \in R$ using four basic components: rules, fuzzifier, inference engine, and defuzzifier. Here, R^m refers to the space of m real numbers and assumes that m measurement deltas are present as inputs to the fuzzy system. The notation R refers to the space of real numbers and refers to the output of the fuzzy system. Thus the fuzzy system performs the mapping

$$F : V \in R^m \rightarrow W \in R$$

$$\text{where } V = V_1 \times V_2 \times \cdots \times V_n \in R^m$$

is the input space and $W \in R$ is the output space. A typical fuzzy logic system maps crisp inputs to crisp outputs using four basic components: rules, fuzzifier, inference engine, and defuzzifier. Once the rules driving the fuzzy logic system have been fixed, the fuzzy logic system can be expressed as a mapping of inputs to outputs.

Rules can come from experts or can be obtained from numerical data. When the rules come from experts, they can be directly represented as words. For example, an expert may suggest that when oil temperature measured by a sensor is high and vibration level at a particular accelerometer A is high, then there is damage at a location B in the structure. A process of interviewing of experts is often the best way to develop a fuzzy rule base directly from human knowledge which encodes the expert information often used by maintenance engineers. However, for many engineering problems, expert knowledge may not be available about the different conditions of the damaged system. In such cases, a mathematical model of the damaged system is invaluable for creating a rule base linking seeded damages to the damaged indicators. Therefore, the rules for a fuzzy logic system can come either from experts or from a mathematical model, depending on the system under consideration. In either case, engineering rules are expressed as a collection of IF-THEN statements such as "IF u_1 is HIGH, and u_2 is LOW, THEN v is LOW." We can see that some of the rules can come from experts while others can come from models.

For SHM, the designer must select the input damage indicators to the fuzzy system and the outputs of the fuzzy system. Various examples and case studies considered in this book will show some typical input-output sets. To physically understand the fuzzy rules, a damage detection problem for a cantilever beam can be considered where the inputs are the changes in natural frequencies relative to the baseline undamaged beam and the output is the location of the damage. To formulate a fuzzy rule, we need an understanding of

1. Linguistic variables (words) versus numerical values of a variable (e.g., HIGH versus 3.5% change in the fundamental frequency),
2. Quantifying linguistic variables (e.g., ω_1 (fundamental frequency) will have a finite number of linguistic terms associated with it, ranging from NEGLIGIBLE, SMALL, MEDIUM, to VERY HIGH), which is done using fuzzy membership functions,
3. Logical connections between linguistic variables (e.g., AND, OR etc.), and
4. Implications such as IF A THEN B. We also need to understand how to combine more than one rule.

Once the inputs of the fuzzy logic system are identified, a set of linguistic variables must be associated with each input. For instance, the change in the first natural frequency may vary from 0–5%. This set of $(0, 5)\%$ is now expressed as words, i.e., (negligible, small, medium, high, very high) assuming a discretization of five levels. This process of moving from the number space to the word space is accomplished by the fuzzification process. The fuzzifier performs the fuzzification and maps numbers into fuzzy sets. Thus a 0.5% change in the input may become a "small" change, and a 4.5% change in the input may become a "very high" change. This number-to-word transformation is very important in fuzzy logic, as all further operations such as rules are performed on the words. Thus, the fuzzifier is needed to activate rules that are expressed in terms of linguistic variables. An inference engine of the fuzzy logic system maps fuzzy sets to fuzzy sets and determines the way in which the fuzzy sets are combined. The inference engine therefore performs the operations for the fuzzy rules by converting the inputs expressed in words to the output expressed in words. The application of the rules on the words rather than on the numbers is the main source of the strength of the fuzzy logic system. The words are relatively insensitive to small changes in numbers and therefore are robust to uncertainty in the inputs. In several applications, numbers are needed as an output of the fuzzy logic system. In those cases, a defuzzifier is used to calculate crisp values from fuzzy values. Thus, the defuzzifier converts the words back to numbers. The internal architecture of the fuzzy system thus operates with words, but the interface to the outside world can be through numbers because of the fuzzifier and the defuzzifier. We will now formally define the various fuzzy logic terms.

Fuzzy Sets A fuzzy set F is defined on a universe of discourse U and is characterized by a degree of membership $\mu(x)$, which can take values between 0 and 1. A fuzzy set generalizes the concept of an ordinary set whose membership function only takes two values, zero and unity. Thus, an element must either belong to an ordinary set or not belong to it. However, an element can belong to two or more fuzzy sets with different degrees of memberships. For example, if we classify height as "tall," "normal," and "short," then a given person may have a high degree of membership in "tall," a low degree of membership in "normal," and a very low degree of membership in "short." This allows the system to deal with ambiguity. Though the concept of height of a person is associated with a clear numerical value, human decisions which are made with height as an input typically use height as a fuzzy variable.

Linguistic Variables A linguistic variable u is used to represent the numerical value x, where x is an element of U. A linguistic variable is usually decomposed into a set of terms $T(u)$, which cover its universe of discourse.

Membership Functions The most commonly used shapes for membership functions $\mu(x)$ are triangular, trapezoidal, piecewise linear, or Gaussian. The designer selects the type of membership function used. There is no theoretical requirement that membership functions overlap. However, one of the major strengths of fuzzy logic is that membership functions can overlap. Fuzzy logic systems are robust because decisions are distributed over more than one input class. For convenience, membership functions are normalized to one so they take values between 0 and 1, and thus define the fuzzy set. One advantage of the Gaussian membership functions is that they do not suddenly go to zero. This helps in a progressive degradation of the behavior of the fuzzy logic system.

Inference Engine Rules for the fuzzy system can be expressed as

$$R_i: \text{IF } x_1 \text{ is } F_1 \text{ AND } x_2 \text{ is } F_2 \text{ AND } \cdots \text{ AND } x_m \text{ is } F_m \text{ THEN } y = C_i$$
$$i = 1, 2, 3, \ldots, M$$

where m and M are the number of input variables and rules, x_i and y are the input and output variables, and $F_i \in V_i$ and $C_i \in W$ are fuzzy sets characterized by membership functions $\mu_{F_i}(x)$ and $\mu_{C_i}(x)$, respectively. Each rule can be viewed as a fuzzy implication

$$F_{12\cdots m} = F_1 \times F_2 \times \cdots \times F_m \rightarrow C_i$$

which is a fuzzy set in $V \times W = V_1 \times V_2 \times \cdots \times V_m \times W$ with membership function given by

$$\mu_{R_i}(x, y) = \mu_{F_i}(x_1) * \mu_{F_2}(x_2) * \cdots * \mu_{F_m}(x_m) * \mu_{C_i}(y)$$

where the asterisk can be the min or product operator with $x = [x_1, x_2, \ldots, x_m] \in V$ and $y \in W$. This sort of rule covers many applications. The algebraic product is one of the most widely used operators in applications and leads to product implication. Underlying all this mathematics is a very simple concept that the degrees of memberships of the different parts of the rule are multiplied to get the degree to which a rule has fired. The Gaussian fuzzy sets are useful here, as they ensure that each rule is fired to some degree as none of the membership functions will become exactly zero. In pattern recognition problems, the outputs are often crisp sets, and $\mu_{C_i}(y) = 1$ is often used for the product implication formula. In this book, product implication will be used for the fuzzy systems.

Defuzzification Popular defuzzification methods include maximum matching and centroid defuzzification. While centroid defuzzification is widely used for fuzzy control problems where a numerical output is needed, maximum matching is often used for pattern matching problems where we need to know the output class. For example, a fuzzy control application may require an output in terms of an angle (degree or radian) for a given actuator. In these applications, the defuzzifier plays an important role. However, in SHM, it may be better to know that the damage is in the "outboard" section of the beam which could have been defined at the region between 60 and 80% of the beam length during the data gathering process. There is no great value in knowing that the damage is at an exact numerical location such as 73% of the beam length, as such a diagnosis is likely to be erroneous given the uncertainties in the problem. The output in words given by the fuzzy system is also useful for maintenance engineers and can be easily fed directly into the graphical user output.

Suppose there are K fuzzy rules and among them, K_j rules ($j = 1, 2, \ldots, L$ and L is the number of classes) produce class C_j. Let D_p^i be the measurements of how the pth pattern matched the antecedent conditions (IF part) of the ith rule, which is given by the product of membership grades of the pattern in the regions which the ith rule occupies:

$$D_p^i = \prod_{i=1}^{m} \mu_{li}, \tag{2.1}$$

where m is the number of inputs and μ_{li} is the degree of membership of measurement l in the fuzzy regions that the ith rule occupies. Let $D_p^{\max}(C_j)$ be the maximum matching degree of the rules (rules $j_l, l = 1, 2, \ldots, K_j$) generating class C_j:

$$D_p^{\max}(C_j) = \max_{l=1}^{K_j} D_p^{jl}. \tag{2.2}$$

Then the system will output class C_j* provided that

$$D_p^{\max}(C_j*) = \max_j D_p^{\max}(C_j). \tag{2.3}$$

If there are two or more classes which achieve the maximum matching degree, we will select the class which has the largest number of fired fuzzy rules (a fired rule has a matching degree of greater than zero).

There are several applications of fuzzy logic systems in SHM as well as in the broader area of engineering. Despite their considerable success, fuzzy systems are limited to problems with a small number of input variables. In addition, the process of developing the fuzzy system requires a lot of judgement and experience on the part of the designer. Two aspects in the design of the fuzzy system are particularly difficult: (1) generating the best rule set and (2) tuning the membership functions. The rules and the membership functions must accurately capture the relationship between the independent and dependent variables.

Unfortunately, the tasks of tuning the membership function and generating rules are not independent. The task of selecting membership functions and rule values is difficult since the information has to be obtained from numerical data of the system to be modeled. Another problem is selecting an appropriate number of fuzzy sets. Most studies use experience to come up with this number. Often, symmetric fuzzy sets are assumed. However, assuming symmetry in the fuzzy sets also implies assuming symmetry in the system being modeled [4]. The results of successful fuzzy logic systems which one reads in papers and which have been implemented in many practical systems have come after much trial and error on the part of the designer. Typically, the designer selects a level of discretization for the measurement, then assigns the membership function for each fuzzy set, creates the rules, and checks for performance. If the performance of the fuzzy system is not good, the level of discretization, membership functions, and rules are manually tuned until a reasonably good level of performance is obtained. However, the danger of this approach is that it is ad hoc in nature and the fuzzy systems developed using this method are not optimal.

To use the power of fuzzy logic for realistic health monitoring problems, it is necessary to automate the process of fuzzy rule creation. For SHM problems, a clear metric for maximization is the success rate of the fuzzy system when confronted with test data. The design of the best fuzzy system for SHM is essentially an optimization problem which involves maximization of the success rate. As discussed in the previous chapter, the process of designing the best fuzzy logic system can be accelerated by using genetic algorithms. We therefore discuss genetic algorithms in the next section.

2.2 Genetic Algorithms

The genetic algorithm (GA) is a search process based on the laws of natural selection and genetics. The GA was one of the early examples of bio-inspiration in engineering and has paved the way for many other such methods and concepts. The GA searches the design space for an optimal design point to maximize a fitness function value. Generally, a simple GA contains three basic operations: selection, genetic operations, and replacement. A typical GA cycle is shown in Fig. 2.2. If Fig. 2.2 is examined, it will be seen that the first step needed to start the GA process is the construction of an initial population. The initial population is the potential solution set comprising a reasonably large number of points in the design space and is generated randomly or heuristically. The general method used at this stage is a random construction of individuals. It is also possible to use a combination of uniform sampling of the design space and random sampling in order to ensure that each part of the design space is represented in the initial population.

The next step of the pre-evolution phase is to evaluate the initial individuals. This step is needed to determine the next generation that will constitute the subpopulation. For a typical problem, the fitness function value needs to be calculated at each of the points in the population. There is a very small but finite probability that the initial population may possibly include the solution.

After this pre-evolution stage, the evolution phase loops until (1) a solution is found or (2) the generation number reaches the predetermined maximum generation

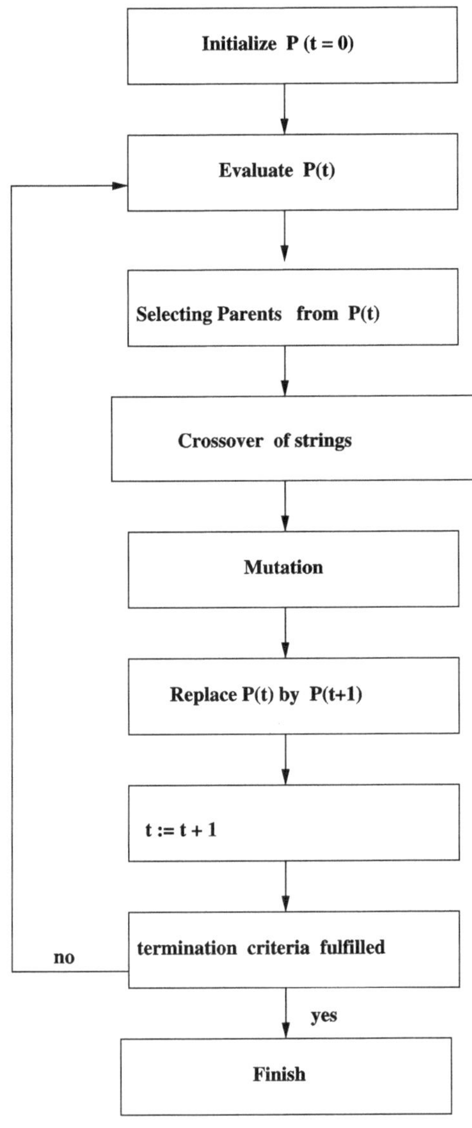

Fig. 2.2 Flowchart of genetic algorithm

number or (3) there remains no mutation combination left to try and any increase in the average success of the population cannot be achieved. Details about GAs are available in textbooks [5, 6], and their use in genetic fuzzy systems is described in [7]. GAs are theoretically and empirically proven to provide a robust search in complex spaces [8]. A GA operates on a population of randomly generated points (P). Each point is sometimes called a chromosome and is often represented by binary strings. There exist both binary and real coded versions of GAs; however, in this book we will use the binary GA, as this works quite well when a low level of discretization in terms of the numerics are needed. In a binary GA, any numbers are first converted into binary form and then the genetic operators are applied to the resulting strings of 1's and 0's. The binary form resulting after the operations can then again be converted to real number form. However, the binary form requires a specification of the number of bits used for representing each design variable. These bits can be kept to a low number for SHM applications in genetic fuzzy systems as they typically represent the characteristics of the fuzzy set.

The GA is an optimization algorithm, and its advantage relative to traditional gradient-based algorithms lies in its ability to locate the global minimum and also operate with discrete or integer design variables. Several terms are widely used in the GA literature, and they are discussed next.

2.2.1 *Operations During a GA Process*

Encoding Encoding is the first part of a GA process, because problem-related information is encoded into a structure called a chromosome or string. A chromosome is generally a sequence of variables of a problem placed in an organized manner. Every variable sequenced to construct the chromosome is called a gene. These definitions come from the biologically inspired nature of the GA. However, as the GA moved away from its biological roots, genes were replaced by bits and chromosomes by strings. For instance, a design variable x can be represented by the string 110110. If there are two such design variables, they can be put side by side and result in a string of double the size. This process of moving from the real number space to the binary space is called encoding.

Fitness Evaluation GAs mimic the survival-of-the-fittest principle of nature to perform a search process. Therefore, GAs are naturally suitable for solving maximization problems where a fitness function is maximized. Minimization problems are usually transformed to maximization problems by some suitable transformation. In general, a fitness function $F(x)$ is first derived from the objective function $f(x)$ and used in successive genetic operations. For maximization problems, the fitness function can be considered to be the same as the objective function, i.e., $F(x) = f(x)$. For minimization problems, the fitness function is an equivalent maximization problem chosen such that the optimum point remains unchanged. A number of such transformations are possible. The following fitness function is often used:

$$F(x) = 1/\big(1 + f(x)\big).$$

GA operators typically require the function value to remain positive. Therefore, the process of changing the sign of the fitness function which is popular in gradient-based optimization is not used in GAs. If there is some chance that the fitness function may become negative in the design space, a large positive number can be added to it to ensure that the fitness stays positive.

Genetic Operations The operation of GAs begins with a population of random strings representing design variables. Thereafter, each string is evaluated to find the fitness value. The population is then operated by three main operators, *reproduction*, *crossover*, and *mutation*, to create a new population of points. These operators are described below.

Reproduction: Reproduction is the first operator applied on a population. Reproduction selects good strings in a population and forms a mating pool. In doing so, it mimics the courtship phase of natural selection. There exist a number of reproduction operators in the GA literature, but the essential idea in all of them is that above-average strings are picked from the current population and their multiple copies are inserted in the mating pool in a probabilistic manner. The commonly used reproduction operator is proportionate reproduction, where a string is selected for the mating pool with a probability proportional to its fitness. This approach is also known as roulette wheel selection. Thus, good strings in a population are probabilistically assigned a larger number of copies and a mating pool is formed. Another approach to reproduction occurs in tournament selection. Here tournaments are arranged between any two random strings, and the winners are selected for mating. It is important to note that no new strings are formed in the reproduction phase.

Crossover: At the end of the reproduction process, the mating pairs are selected. The idea of exchange of genetic information which occurs between the male and female is now mimicked during the crossover process. Recall that each individual is a point or a binary string. Thus, information exchange involves the swapping of some bits between the male and female strings. The process of information exchange between the individuals is called crossover and is a basic property of GAs. The crossover procedure creates new chromosomes or strings from the two parents. Crossover is performed after selection of a subpopulation of individuals according to their fitness values and collection of the selected individuals into a gene pool. Crossover is achieved in three stages. The first stage is matching. Matching is the selection of two individuals in the gene pool randomly. In the second stage, a crossover point is determined in each of the individuals. In the final stage, two parts of the individuals are replaced with each other. Typically, the crossover performs an operation where two parents lead to two children. An example of crossover is shown below:

Parent Strings		Children Strings
100 | 1001		100 | 0111
	==Crossover⟹	
111 | 0111		111 | 1001

Besides the single-point crossover shown above, multipoint crossover can also be used. In such a case, the parent strings between two sites are swapped to get the child strings.

Parent Strings		Children Strings
100 | 1001 | 1001001		100 | 0111 | 1001001
	==Crossover⟹	
111 | 0111 | 0101010		111 | 1001 | 0101010

Multipoint crossover is better when a large number of design variables result in long strings. For such long strings, single-point crossover is biased toward the right of the strings and changes these design variables much more often than those at the left of the string.

Mutation: Mutation means a random change in the information of a chromosome or string. In other words, mutation is an operation that defines the variation in a chromosome. This variation may be local or global. A probability test determines whether a mutation will be carried out or not. For example, if the average fitness of the new generation is smaller than the average fitness of the previous generation, bit y of the chromosome x can be changed. A bit mutation applied to a chromosome is shown below:

Before mutation: 100 | 1 | 1101

After mutation: 100 | 0 | 1101

Mutation can be useful for improving the population. Since the initial population is a subset of all possible solutions, an important bit of all the chromosomes may be 0 while it must be 1 to be optimal. Crossover may not solve this problem and mutation is indispensable for the solution. In general, mutation leads to small local moves in the design space, while crossover leads to larger global moves. Mutation may often lower the average fitness of the population, but it also allows the GA to escape from a local minimum by adding diversity.

There are two proposed methods for allowing a subpopulation to replace its ancestors. One of the methods is generational replacement. In this method, a population of size n entirely replaces the new generation. The other method is steady-state reproduction, which replaces only a few individuals in a generation. A small number of strings with high fitness values are sometimes shielded from the crossover and mutation operations. This is known as elitism.

2.2.2 *Performance Factors*

Population size is a factor that affects the performance of the GA. Increasing the population means a longer computation time. On the other hand, if the population size is decreased, the accuracy of the solution also decreases because of reduced variation of chromosomes. In GA design, there must be a balance between the generation numbers and population size. Population size has another effect in the GA; it reduces the effect of the highest fitness valued chromosomes. For example, in a population of 10 chromosomes, if one of the chromosomes has a fitness value of 9 while the others have a fitness value of 1, half of the parents will be chosen from among the relatively low fitness valued chromosomes, although the best fitness valued chromosome is nine times better. Evaluation of chromosomes and fitness calculations are the most time-consuming parts of the GA. If the evaluation operation is reduced, the GA process will work faster, and this can be achieved by reducing the population size and number of generations needed to reach a solution.

2.3 Genetic Fuzzy System

The principles and operations of GAs and fuzzy logic have been briefly described in the previous sections. These two soft computing tools can be combined to form the genetic fuzzy system [9]. The GA provides good global search capability. Fuzzy logic presents robust and flexible inference methods in problems subject to imprecision and uncertainty. The linguistic representation of knowledge permits a person to interact with a fuzzy system in an easy manner. The hybridization of the GA and fuzzy logic gives an advanced soft computing method called the genetic fuzzy system (GFS), in which a GA is used to evolve a fuzzy system by tuning fuzzy membership functions and learning fuzzy rules.

A key objective of this section is to use GAs to automate the design of fuzzy systems. The generalized GFS algorithm is explained for a damage detection problem.

Input and Output Suppose inputs to the fuzzy systems are represented by **z** and outputs are represented by **x**. The objective is to find the mapping between **z** and **x**. In damage detection problems, the measurement deltas Δz's (changes in measurements between damaged and undamaged model) can be used as inputs, and the combinations of damage levels and locations will be the output of the fuzzy system. Here $\mathbf{x} = \{location_1, location_2, \ldots, location_n\}^T$ and $\mathbf{z} = \{\Delta z_1, \Delta z_2, \ldots, \Delta z_d\}^T$, where n are the user-defined locations and d is a user-defined number of measurements. Each measurement delta has uncertainty.

Fuzzification The structure can be divided into various locations. For example, "$location_1$" ranges from 0% to say n_1% and so on until all the locations are labeled per user requirement. To get a degree of resolution of the extent of damage [10], each of these damage locations is allowed several levels of damage and split into linguistic variables. For example, consider "$location_1$" as a linguistic variable. Then it can be decomposed into k levels and the undamaged level,

$$T(location_1) = \{Undamaged, level_1\ Damage, level_2\ Damage, \ldots, level_k\ Damage\}$$

where each term in $T(location_1)$ is characterized by a fuzzy set in the universe of discourse $U(location_1) = \{0, damage\ parameter\}$. The other structural damage variables are fuzzified in a similar manner.

The measurement deltas $\Delta z_1, \Delta z_2, \ldots$ and Δz_d are also treated as fuzzy variables. Fuzzy sets with Gaussian membership functions are used to define these input variables. These fuzzy sets can be defined using the following equation:

$$\mu(x) = e^{-0.5(\frac{x-m}{\sigma})^2}, \tag{2.4}$$

where m is the midpoint of the fuzzy set.

Δz's are used as midpoints for the respective fuzzy sets. The standard deviation associated with each variable is denoted by σ and is obtained using a GA for maximization of the success rate S_R, as discussed later. After optimization of the fuzzy system for its success rate, we get a different σ value for every fuzzy set. We can see that the Gaussian fuzzy sets depend on the appropriate choice of the midpoint and the standard deviation. The midpoint is a measure of the point of maximum likelihood of a fuzzy set, while the standard deviation represents the scatter and accounts for the uncertainty. For instance, we could define a 0% change in the natural frequency as "negligible" and give a standard deviation of 0.3%. This ensures that a Gaussian fuzzy set for the word "negligible" is centered at 0 change in frequency representing the undamaged condition and also that very small shifts from the 0 position are also classified as negligible. Thus, the numerical value is fuzzified by the spread of the Gaussian fuzzy set, and the standard deviation allows us to control how slowly or fast the "negligible" membership function decays.

Rule Generation Rules for the fuzzy system can be obtained by fuzzification of the numerical values obtained by numerical analysis of the model using the following procedure.

1. The fuzzy sets corresponding to $\Delta z_1, \Delta z_2, \ldots, \Delta z_d$ are generated by taking the Δz's obtained by numerical analysis as midpoints of membership functions corresponding to a given structural damage. This strategy for selecting the midpoint ensures that the maximum degree of membership ($\mu = 1$) for each fuzzy set occurs at the values of Δz since the Gaussian function is highest at the midpoint. The standard deviation of each set is initially fixed randomly within a prescribed range.
2. For each measurement delta corresponding to given fault, the degree of membership in the fuzzy set is calculated.
3. Each measurement delta is assigned to the fuzzy set with the maximum degree of membership. One rule is obtained for each damage type by relating the measurement deltas. The "IF" rule segment before "THEN" part will change from structure to structure but the rule segment after "THEN" will remain the same unless the number of damage locations and levels of damages are changed. Suppose for rule i we are giving d inputs. Then the fuzzy system will generate d membership functions A taking the change in measurements obtained by numerical analysis of the model as the midpoint. The standard deviation is obtained by optimization as discussed later.
4. The fuzzy system is then fine tuned by changing the standard deviation σ for each set.

Table 2.1 Rules for fuzzy system

Rule	Δz_1	Δz_2	...	Δz_d
Undamaged	A_{11}	A_{12}	...	A_{1d}
$level_1$ Damage at $location_1$	A_{21}	A_{22}	...	A_{2d}
$level_1$ Damage at $location_2$	A_{31}	A_{32}	...	A_{3d}
$\vdots$				
$level_1$ Damage at $location_k$	$A_{(k+1)1}$	$A_{(k+1)2}$	...	$A_{(k+1)d}$
$\vdots$				
$level_n$ Damage at $location_k$	$A_{((n\times k)+1)1}$	$A_{((n\times k)+1)2}$	...	$A_{((n\times k)+1)d}$

So rule i will be in generalized form as:

$$\begin{aligned}&\text{IF}\\&\quad \Delta z_1 \text{ is } A_{i1} \text{ AND}\\&\quad \Delta z_2 \text{ is } A_{i2} \text{ AND}\\&\quad \vdots\\&\quad \Delta z_d \text{ is } A_{id} \text{ AND}\\&\text{THEN}\end{aligned}$$

corresponding level of damage at corresponding location.

In the preceding rules, the membership values of membership function A will change from structure to structure. These rules are symbolically tabulated in Table 2.1.

For calculation of the uncertainty associated with variables, i.e., the standard deviation of the Gaussian membership functions, we use a GA for optimization of the success rate. As we have already discussed, there will be uncertainty and some noise in the measurement deltas. By generating noisy deltas and testing the fuzzy system for a known damage, we can define a success rate. Our optimization problem can be written in standard form as

$$\begin{aligned}\text{Maximize}\quad & S_R\\ \text{For design variable}\quad & \sigma^{\min} \le \sigma_{ij} \le \sigma^{\max}\\ & \text{where } i = 1, 2, \ldots, M,\\ & \qquad\quad j = 1, 2, \ldots, d,\end{aligned}$$

where M is the number of rules and d is the number of fuzzy sets.

The success rate is calculated using the results obtained after defuzzification. If we test N_T samples of noisy data and out of that the system correctly classifies N_C times, then the success rate for rule p as a percentage is given as [11]

$$S_R = \frac{N_C}{N_T} 100. \tag{2.5}$$

The added noise in the data simulates the uncertainty present in the experimental measurements and the modeling process. Given a computed measurement delta Δz, random number u in the interval $[-1, 1]$, and a noise level parameter α, the noisy simulated data is given as

$$\Delta z^{\text{noisy}} = \Delta z + u\alpha. \tag{2.6}$$

The parameter α defines the maximum variance between the computed value of Δz and simulated measured value Δz^{noisy} which is a simulation of a practical measurement. For example, if $\alpha = 0.15$, then the simulated measurement delta can be different by as much as 0.15 from the ideal value predicted by simulations. Thus, α can be used to control noise levels in the simulated data used for testing the fuzzy logic system.

2.4 Summary

This chapter gives a brief introduction to fuzzy logic, genetic algorithms, and a few other terms used in the formulation of the genetic fuzzy system. Fuzzy logic involves computing with words instead of numbers and is robust to the presence of uncertainty in the inputs. However, fuzzy logic systems are difficult to design due to the interaction between the choice of appropriate membership functions, the rule base obtained, and the performance of the fuzzy system. The process of designing fuzzy systems can be automated by using an optimization procedure for tuning the membership functions and rule base. Genetic algorithms provide an excellent approach for finding the global minimum and can be used to design the fuzzy systems. The genetic fuzzy system combines the uncertainty representation characteristics of fuzzy logic with the learning ability of the genetic algorithm. Using the changes in measurement deltas, a fuzzy system is generated, and the rule base and membership functions are optimized by the genetic algorithm. The generalized formulation of the genetic fuzzy system is explained using a damage detection problem. The genetic fuzzy system will be used for solving damage detection problems using modal data in the next two chapters. These applications will make the process of developing genetic fuzzy systems for structural health monitoring very clear.

References

1. Kosko, B.: Fuzzy Engineering. Prentice-Hall, Englewood Cliffs (1997)
2. Ross, T.J.: Engineering Applications of Fuzzy Logic. Wiley, New York (2010)
3. Tanaka, K., Nimura, T.: An Introduction to Fuzzy Logic for Practical Applications. Springer, Heidelberg (1996)
4. Karr, C.L., Stanley, D.A., McWhorter, B.: Optimization of hydrocyclone operation using a geno-fuzzy algorithm. Comput. Methods Appl. Mech. Eng. **186**(2–4), 517–530 (2000)
5. Goldberg, D.: Genetic Algorithms in Search, Optimization, and Machine Learning. Pearson Education, Boston (2001)

6. Haupt, R.L., Haupt, S.E.: Practical Genetic Algorithms. Wiley Interscience, New York (2004)
7. Cordon, O., Herrera, F., Villar, P.: Generating the knowledge base of a fuzzy rule-based system by the genetic learning of the data base. IEEE Trans. Fuzzy Syst. **9**(4), 667–674 (2001)
8. Coello, C.A.C.: Theoretical and numerical constraint handling techniques used with evolutionary algorithms: a survey of the state-of-the art. Comput. Methods Appl. Mech. Eng. **191**, 1245–1287 (2002)
9. Cordon, O., Harrera, F., Hoffman, F., Magdalena, L.: Genetic Fuzzy Systems. World Scientific, Singapore (2002)
10. Ganguli, R.: A fuzzy logic system for ground based structural health monitoring of a helicopter rotor using modal data. J. Intell. Mater. Syst. Struct. **12**(6), 397–408 (2001)
11. Abe, S.: Pattern Classification: Neuro-Fuzzy Methods and Their Comparison. Springer, London (2001)

Chapter 3
Structural Health Monitoring of Beams

Structural damage detection is an inverse problem of structural engineering having four main parts: finding the existence, location, and extent of damage, and suggesting a course of remedial action often called prognosis. In this chapter, a genetic fuzzy system (GFS) is used to find the location and extent of damage in a beam. The beam is a fundamental structural element used to model systems such as helicopter rotor blades, airplane wings, columns, bridges, and buildings. In fact, the beam is probably the most ubiquitous structural member. The GFS automatically generates the rules for a fuzzy system using a genetic algorithm (GA) for application to structural damage detection. The GFS is demonstrated for damage detection in a beam-type structure modeled using the finite element method. The finite element method is a numerical method for solving governing differential equations and has proved to be particularly successful for problems in solid mechanics where nonuniformity of the domain and boundary conditions do not permit analytical solutions.

We first consider a cantilever beam for illustrating the GFS for the damage detection problem. Cantilever beam-type structures include, e.g., helicopter blades, turbine and compressor blades, wind turbine blades, and airplane wings. These structures are fixed at one end and free at the other end. We also discuss results for the BO105 hingeless helicopter rotor blade at the end of this chapter.

In the present work, the fuzzy rules are automatically generated. The GA is used to solve the optimization problem. It is based on roulette wheel selection, fixed-point crossover, and bitwise mutation. The population size, crossover probability, mutation probability, and maximum number of generations are 20, 0.8, 0.05, and 30, respectively. These values are determined by numerical experimentation. It should be noted that the computational expense of the GA here depends on the underlying finite element analysis. For the beam-type structures considered in this chapter, the number of degrees of freedom is small, and hence the computer time for solving the eigenvalue problem is much less. All numerical results in this book are obtained on PCs, reflecting the growing power of computers.

In this chapter, a GFS is developed and demonstrated for damage detection in two isotropic beam-type structures using modal-based measurement deltas. A cantilever beam and a hingeless helicopter blade with only flapwise (out-of-plane) bending are considered.

P.M. Pawar, R. Ganguli, *Structural Health Monitoring Using Genetic Fuzzy Systems*,
DOI 10.1007/978-0-85729-907-9_3,

3.1 Damage Modeling in 1D Beam

The GFS is demonstrated using simulated signals of a one-dimensional (1D) isotropic cantilever beam which is modeled using the Euler–Bernoulli cantilever beam with the governing equation

$$\frac{\partial^2}{\partial x^2}\left[EI(x)\frac{\partial^2 w}{\partial x^2}\right]+m(x)\frac{\partial^2 w}{dt^2}=0. \tag{3.1}$$

Here $EI(x)$ is the flexural stiffness, $m(x)$ is the mass per unit length, and $w(x,t)$ is the displacement. To obtain the natural frequencies, harmonic excitation is assumed: $w(x,t)=W(x)e^{i\omega t}$. This converts the partial differential equation (PDE) into an ordinary differential equation (ODE). This ODE has an exact solution for natural frequencies for a uniform beam which has constant flexural stiffness and mass per unit length. However, structural damage causes a localized reduction in the flexural stiffness at the point where the damage occurs. Therefore, modeling of damaged beams requires analysis of a nonuniform beam.

For a nonuniform beam, an approximate method such as the finite element method is needed to calculate the natural frequencies. In finite element analysis, the structure is divided into many small finite elements and a local displacement field is assumed within the element with degrees of freedom present at the element nodes. The elements are then assembled together and appropriate boundary conditions are applied. Basically, finite element analysis transforms a differential equation into a matrix equation and gives a discrete system approximation of the continuous system. Details of the finite element method are given in standard textbooks [1–4]. The beam element used here has four degrees of freedom, two at each node. The nodal degrees of freedom are the displacement and slope at the two ends of the element. Between the finite elements, there is continuity of the displacement and slope degrees of freedom. The element mass and stiffness matrices for this beam element are of size 4×4. The element matrices for the entire beam are assembled and boundary conditions are enforced to give the global stiffness matrix $\mathbf{K}_g$ and the global mass matrix $\mathbf{M}_g$. For a cantilever beam which is hinged at the root, the displacement and slope degree of freedom at the root node is set to zero. These are known as geometric boundary conditions for the cantilever beam. Then, for free vibration, we have the eigenvalue problem as

$$\mathbf{K}_g\Phi=\omega^2\mathbf{M}_g\Phi \tag{3.2}$$

This equation is solved numerically as a general eigenvalue problem to obtain the natural frequencies ω and the eigenvectors Φ of the beam. Such equations can easily be solved in MATLAB, for instance, or by using any other numerical computing software.

The change in frequency is considered as the measurement delta and is calculated by the finite element method for a combination of five different locations and four different levels of damages: Undamaged, Slight Damage, Moderate Damage, and

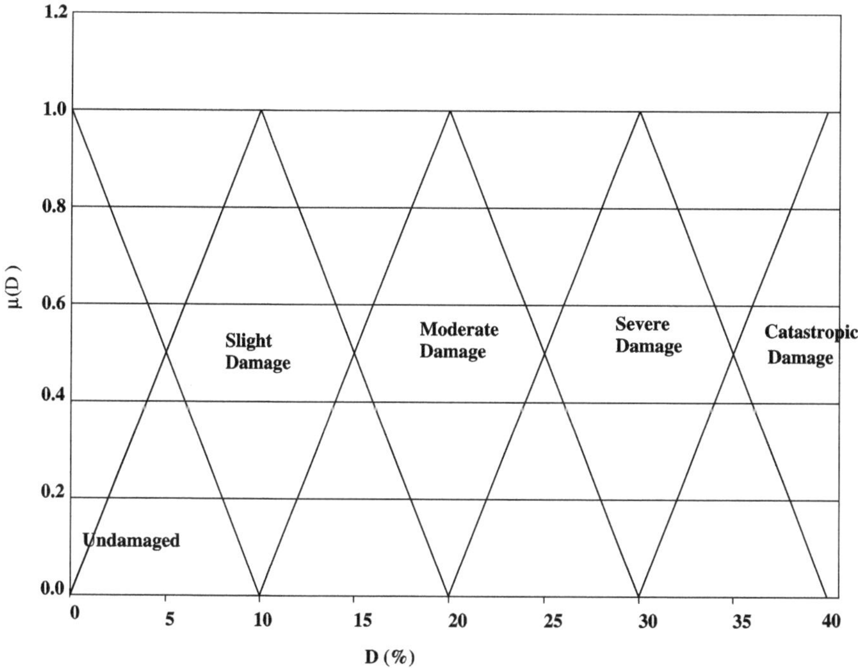

Fig. 3.1 Fuzzy sets representing damage levels over universe of discourse (0–40%)

Severe Damage. The damage is modeled as a reduction in element stiffness, and a percentage damage parameter D is defined such that [5–7]

$$D = 100\frac{E_{\text{undamaged}} - E_{\text{damaged}}}{E_{\text{undamaged}}}. \tag{3.3}$$

Here E is the Young's modulus of the material. The structural damage in each segment is modeled by stiffness reductions (D) of 10%, 20%, and 30%, respectively. These damage sizes are classified as "slight damage," "moderate damage," and "severe damage," respectively. Damage sizes below "slight damage" are classified as undamaged. Damage sizes greater than "severe damage" are classified as "catastrophic damage" as shown in Fig. 3.1. Thus both the damage locations and the damage size are expressed in terms of words to make the problem suitable for fuzzy logic.

The beam is divided into five damage locations as "root" ranging from 0–20% of the beam, "inboard" from 20–40%, "center" from 40–60%, "outboard" from 60–80%, and "tip" from 80–100%, as shown in Fig. 3.2. Since modal properties such as frequencies are global properties in contrast to strains, which are local properties, it is best to use the modal methods for a broad classification of the damage and then to subject the structure to more detailed local inspection using nondestructive testing methods. The difference between the frequency of the damaged and undamaged beam is used as a system indicator for damage and referred to as

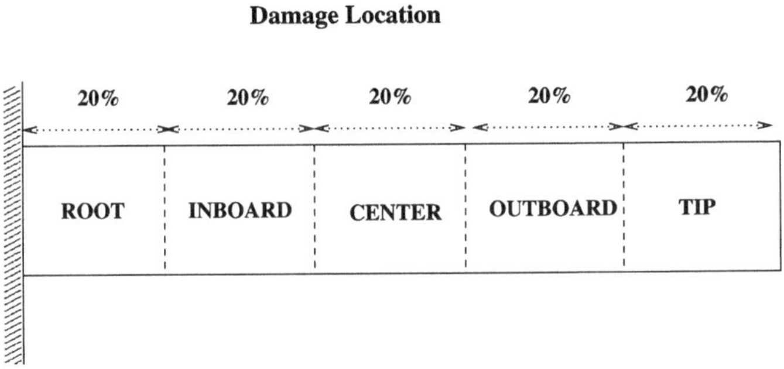

Damage Location

20% 20% 20% 20% 20%

ROOT INBOARD CENTER OUTBOARD TIP

Fig. 3.2 Schematic representation of the cantilever beam damage locations

a "measurement delta." This measurement delta is positive for structural damage since the reduction in stiffness for a damaged beam decreases the frequency. The measurement delta is expressed as a percentage change:

$$\Delta\omega = 100\frac{\omega_{\text{undamaged}} - \omega_{\text{damaged}}}{\omega_{\text{undamaged}}}. \tag{3.4}$$

Different locations of damage in the beam and different levels of the damages give different sets of $\Delta\omega$, which are used to create the database of fuzzy rules. Note that in this problem, the database used to create rules is obtained from a mathematical model. It is also possible to obtain such a database from experiments. However, the costs of performing experiments on the large number of damaged beams can be very high. Therefore, a model-based diagnostics is very useful.

The sensitivity of $\Delta\omega$ with respect to damage level and location is shown in Fig. 3.3. These results are obtained from the finite element model of the cantilever beam by seeding the damage at different locations along the beam. However, any modeling process including the finite element method contains errors due to uncertainty in the material properties and discretization and numerical errors. Uncertainties in the model can be of two types: aleatory and epistemic [8]. Aleatory or random uncertainty can be present because the material properties used in the deterministic mathematical model such as Young's modulus are actually random variables. In other words, if we were to build ten beams and find the Young's modulus for each beam, there would be some scatter in the data obtained. Besides random uncertainty, epistemic or model uncertainty may also be present. For instance, if Euler–Bernoulli beam theory is used to model a short thick beam, the results will not be accurate, as the effects of shear deformation and rotary inertia are neglected in the Euler–Bernoulli beam theory. These effects are included in Timoshenko beam theory, and the use of such a theory for short thick beams will reduce epistemic uncertainty.

While epistemic uncertainty can be reduced by choosing good models, random uncertainty is harder to reduce as it involves manufacturing and quality issues. Discretization errors arise from not using enough elements in the finite element model. It is very important to use a sufficient number of finite elements especially at locations of sudden change in the material properties which occur in the vicinity of the damage. As a rule, it is important to perform a convergence study by increasing

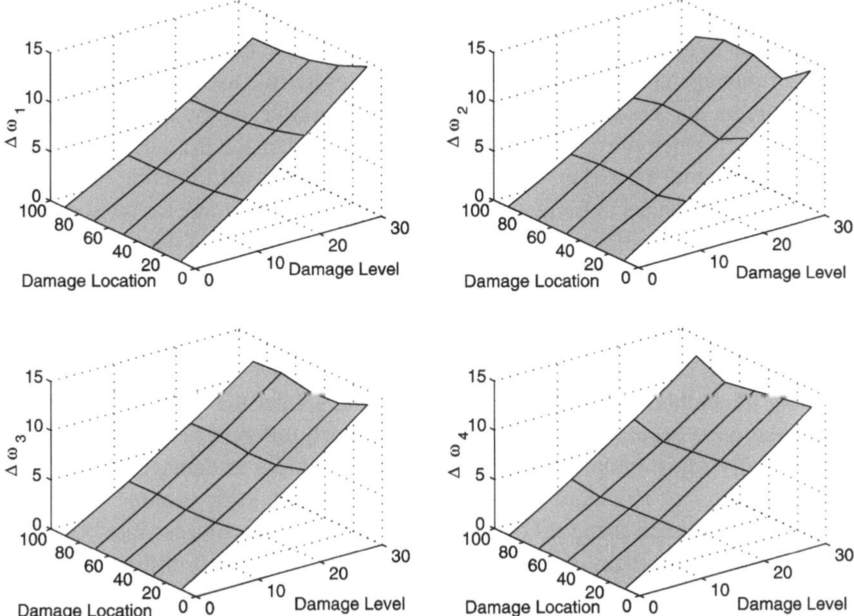

Fig. 3.3 Surface plots of $\Delta\omega$ for first four frequencies with respect to damage location (Root = 0–20%, Inboard = 20–40%, Center = 40–60%, Outboard = 60–80%, Tip = 80–100%) and damage level (undamaged, slight damage, moderate damage, and severe damage at $D = 0\%$, 10%, 20%, and 30%, respectively)

the number of finite elements and monitoring the damage indicator value. It is better to overdiscretize and err on the side of caution than to underdiscretize. We thus see that there is always some difference between predictions by models and test results, even if discretization errors have been minimized. This difference is called modeling uncertainty. In addition to modeling uncertainty, noise may be present in the frequency measurement deltas ($\Delta\omega$'s). This is called measurement uncertainty. In this work, we will assume that random noise models both forms of uncertainties present in the structure. Finally, numerical errors involved in algorithms such as eigenvalue solvers, linear system solvers, and differential equation solvers should be minimized by using robust methods and double precision computer arithmetic.

3.2 Formulation of Genetic Fuzzy System

Formulation of the genetic fuzzy system (GFS) is demonstrated for beam-type structures for a cantilever beam. The inputs of the fuzzy system are $\{\Delta\omega_1, \Delta\omega_2, \ldots, \Delta\omega_d\}^T$ and the outputs of the fuzzy system are various damage levels at five fault locations {Root, Inboard, Center, Outboard, Tip}. The objective of the GFS is to find a mapping between the frequency-based measurement deltas and damage at the five locations. Each measurement delta has uncertainty. Helicopter rotor blades are designed to have a high level of damage tolerance and therefore only significant levels of damage are of interest to us here. Therefore, the use of a global method such as frequency-based damage detection is appropriate, as the damage indicator acts as a filter which prevents minor and insignificant damage from being detected.

Fuzzy sets with Gaussian membership functions are used to define these input variables. The changes in frequencies ($\Delta\omega$) are used as midpoints for the respective fuzzy sets. This midpoint will change for different structures and damage combinations since it is dependent on geometrical properties and material properties of the structure. As this GFS automatically adjusts the midpoints of fuzzy sets, it has the flexibility to deal with different beams. In addition, by selecting the number of fuzzy sets as equal to the number of measurements or modes, we automate the process of selecting the number of fuzzy sets. Of course, the greater the number of measurements, the more accurate the fuzzy system. For each different beam, thc GFS will automatically adjust midpoints according to the conditions of damage and location.

The uncertainty (standard deviation) associated with each variable is denoted by σ and is obtained using a GA for maximization of the success rate S_R. After optimization of the fuzzy system for its success rate, we get a different σ value for every fuzzy set. For calculation of the uncertainty associated with the variables, i.e., the standard deviation of the Gaussian membership functions, we use a GA for optimization of the success rate. As we have already discussed, there will be the presence of uncertainty and some noise in the measurement deltas. The added noise in the data simulates the uncertainty present in the experimental measurements and the modeling process. The data for training and testing of the GFS is developed by using the noise model given in (2.6) by considering the Δz as frequencies obtained by finite element analysis.

The schematic representation of the GFS is shown in Fig. 3.4. Figure 2.1 in the previous chapter shows the components of a classical fuzzy logic system where data is fuzzified, the rules are evaluated, and the defuzzification is used to determine the final output of the system. For our problem, inputs are the measurement deltas ($\Delta\omega$) and the outputs are the damage presence, location, and size. Figure 3.4 shows how the fuzzy logic system in Fig. 2.1 is developed. Data from the finite element model are used to obtain the midpoints of the fuzzy sets. Noisy data at noise level α are used to train the fuzzy system to maximize the success rate (SR) by using the standard deviations as design variables. The outputs of the GFS are damage presence, location, and size, which are the same as those for the fuzzy system in Fig. 2.1. However, the addition of learning using GA allows the GFS to optimize its rule base.

Once we obtain the proper fuzzy rules which are generated automatically, we have a knowledge base to isolate structural damage using frequency shifts. When these rules are applied for a given measurement, we have the degree of membership for each fault. For fault isolation, wc arc intcrcstcd in the most likely fault. Therefore, the fault with the highest degree of membership is selected as the most likely fault. The output is left as a fuzzy set, as it is closer to the human reasoning process. For example, it is better to know that there is a "moderate" damage in the "center"

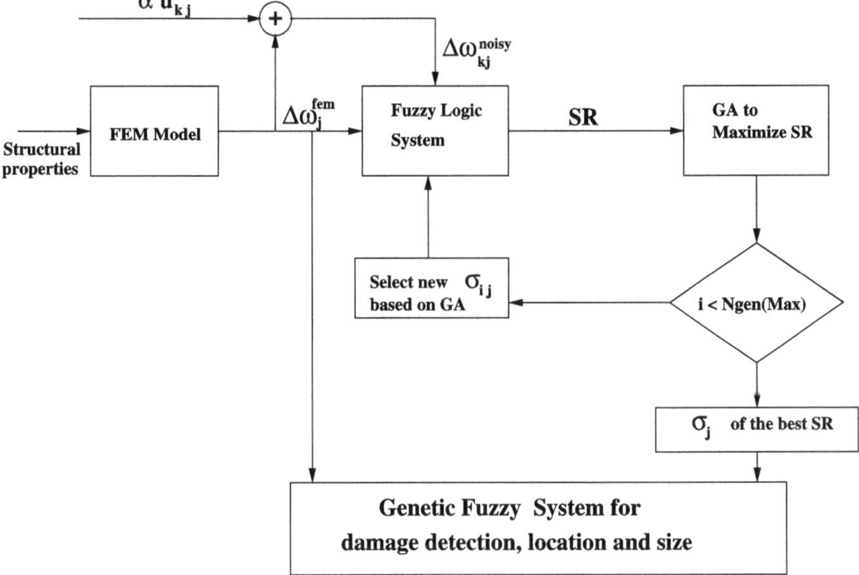

Fig. 3.4 Schematic representation of GFS

element than to know that there is a D of 22% at a distance of $0.46R$ from the root of the beam. The fuzzy output could be color coded or formulated in terms of "alerts" or "alarms" at different levels, which is useful for maintenance actions.

3.3 Structural Health Monitoring for Uniform Beam

The schematic representation of the beam is shown in Fig. 3.2. The dimensions and material properties of the beam (Fig. 3.5) are obtained from Chandrupatla and Belegundu [1]. The beam is divided into 20 finite elements of equal length. The selection of the number of elements is justified in Fig. 3.6 to minimize the modeling error. In this figure, the ratio of the eighth mode, which is the highest mode used in the numerical results, with the first mode is shown. From the graph, it appears that 20 elements give a good level of discretization. Each segment spanning 20% of the beam in Fig. 3.2 is therefore divided into four finite elements. The undamaged beam is uniform. Therefore, the frequency predictions from the finite element method (FEM) model of the undamaged beam are validated by comparing with exact solutions for a continuous beam. The first eight natural frequencies of the beam are 22.66 Hz, 142.03 Hz, 397.70 Hz, 779.37 Hz, 1288.49 Hz, 1925.18 Hz, 2689.82 Hz, and 3583.07 Hz, respectively. We shall assume in this study that the finite element analysis is accurate and any deviations from reality are random noise and not systematic (biased) modeling error. For applications to real structures, some methods such as finite element model updating can be used to match the model predictions of the undamaged structure with experimental data.

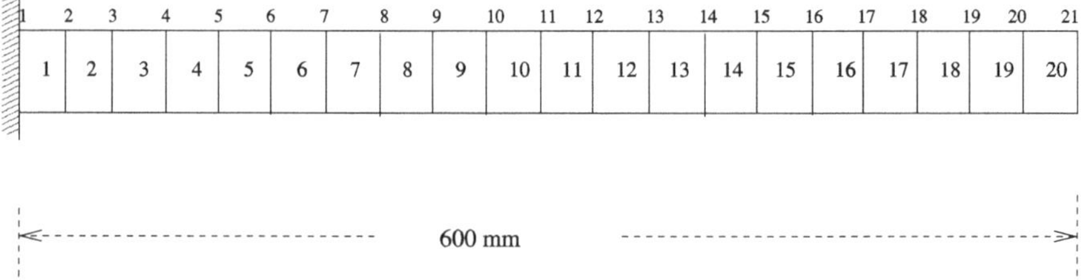

Fig. 3.5 Cantilever beam model for damage detection with 20 finite elements. Material and geometric properties of the beam are: $E = 2.0 \times 10^5$ N/mm^2, cross-sectional area = 240 mm^2, moment of inertia = 2000 mm^4, and mass density = 7840×10^{-9} kg/mm^3

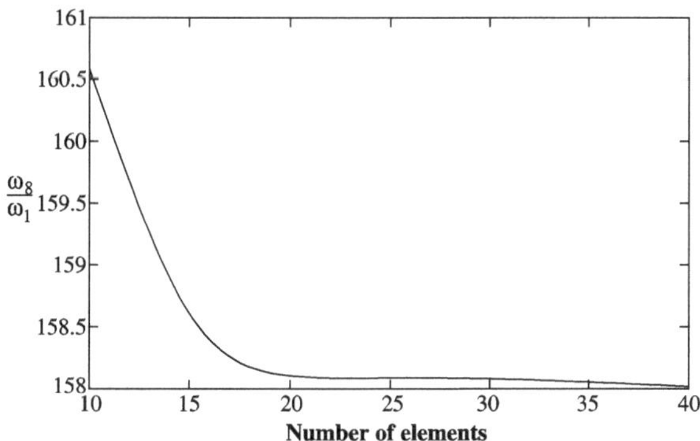

Fig. 3.6 Ratio of eighth frequency to first frequency of the cantilever beam with increasing number of finite elements

The generated rules are shown in Table 3.1 for eight input deltas. The eight input deltas represent the first eight natural frequencies of the beam. During the process of developing the fuzzy system, eight frequencies are used, but during testing, we will also experiment with missing measurements. In the GFS, only the beam geometry and material properties need to be specified along with the number of measurement deltas (modes) needed, and the rule base and success rate for the optimal system are automatically generated. In Table 3.1, the midpoints and standard deviation of membership functions are shown for respective faults in linguistic form, which is the output. The linguistic forms will remain constant for different structures, but the numerical values of midpoints and standard deviation will change. We can see that a lot of information is contained in this table, and it is very difficult to find all this information manually through a trial-and-error process.

3.3.1 *Test with Noisy Data*

The fuzzy logic system is optimized for $\alpha = 0.15$ and also tested using noise contaminated simulated data. The fuzzy system defined by rules for eight $\Delta\omega$ inputs (Table 3.1) is tested for different values of α as shown in Table 3.2. Table 3.2 contains the success rate for each rule for values of α of 0.1, 0.15, and 0.20. This success

Table 3.1 Midpoint and standard deviation of fuzzy sets for different rules ($d = 8$, $\alpha = 0.15$)

Rule No.	Rule	$\Delta\omega_1$	$\Delta\omega_2$	$\Delta\omega_3$	$\Delta\omega_4$	$\Delta\omega_5$	$\Delta\omega_6$	$\Delta\omega_7$	$\Delta\omega_8$
1	Undamaged	0.00 (0.32)	0.00 (0.33)	0.00 (0.30)	0.00 (0.34)	0.00 (0.26)	0.00 (0.30)	0.00 (0.28)	0.00 (0.31)
2	SLD at Root	4.45 (0.34)	4.27 (0.31)	4.13 (0.34)	3.98 (0.31)	3.61 (0.30)	4.55 (0.29)	4.21 (0.33)	4.14 (0.33)
3	SLD at Inboard	4.00 (0.27)	3.50 (0.29)	3.71 (0.33)	3.84 (0.25)	3.33 (0.26)	4.19 (0.31)	4.07 (0.27)	4.28 (0.34)
4	SLD at Center	3.72 (0.28)	3.89 (0.27)	3.61 (0.34)	3.65 (0.27)	3.48 (0.32)	4.22 (0.25)	4.34 (0.25)	3.92 (0.31)
5	SLD at Outboard	3.58 (0.33)	3.92 (0.26)	3.84 (0.30)	3.46 (0.34)	3.56 (0.25)	4.32 (0.30)	4.06 (0.30)	4.27 (0.31)
6	SLD at Tip	3.53 (0.25)	3.56 (0.32)	3.74 (0.29)	3.92 (0.25)	3.63 (0.30)	4.49 (0.26)	3.97 (0.33)	3.92 (0.26)
7	MOD at Root	9.26 (0.32)	8.90 (0.34)	8.62 (0.26)	8.35 (0.26)	7.64 (0.26)	9.40 (0.25)	8.74 (0.29)	8.62 (0.25)
8	MOD at Inboard	8.40 (0.32)	7.40 (0.33)	7.78 (0.33)	8.03 (0.33)	7.03 (0.29)	8.69 (0.31)	8.45 (0.25)	8.87 (0.27)
9	MOD at Center	7.83 (0.29)	8.15 (0.25)	7.61 (0.28)	7.65 (0.32)	7.30 (0.26)	8.75 (0.34)	9.00 (0.26)	8.17 (0.30)
10	MOD at Outboard	7.55 (0.32)	8.22 (0.29)	8.03 (0.28)	7.29 (0.32)	7.43 (0.25)	8.96 (0.26)	8.45 (0.33)	8.84 (0.25)
11	MOD at Tip	7.47 (0.34)	7.51 (0.32)	7.86 (0.28)	8.20 (0.30)	7.56 (0.32)	9.30 (0.26)	8.24 (0.29)	8.17 (0.31)
12	SVD at Root	14.49 (0.33)	13.96 (0.28)	13.56 (0.32)	13.17 (0.34)	12.18 (0.28)	14.62 (0.31)	13.67 (0.31)	13.50 (0.34)
13	SVD at Inboard	13.25 (0.34)	11.77 (0.29)	12.30 (0.25)	12.65 (0.28)	11.16 (0.32)	13.58 (0.25)	13.19 (0.25)	13.82 (0.29)
14	SVD at Center	12.42 (0.32)	12.86 (0.34)	12.08 (0.32)	12.08 (0.30)	11.53 (0.34)	13.65 (0.29)	14.05 (0.27)	12.82 (0.29)
15	SVD at Outboard	12.00 (0.33)	12.98 (0.28)	12.67 (0.28)	11.60 (0.27)	11.70 (0.32)	13.96 (0.26)	13.27 (0.25)	13.78 (0.31)
16	SVD at Tip	11.89 (0.28)	11.94 (0.34)	12.45 (0.25)	12.91 (0.31)	11.87 (0.29)	14.49 (0.27)	12.88 (0.28)	12.81 (0.26)

SLD = Slight damage

MOD = Moderate damage

SVD = Severe damage

Quantities in parentheses represent standard deviation

Table 3.2 Success rate for different rules at different noise levels ($d = 8$)

Rule No.	$\alpha = 0.1$	$\alpha = 0.15$	$\alpha = 0.2$
1	100.00	100.00	100.00
2	100.00	100.00	100.00
3	100.00	100.00	100.00
4	100.00	100.00	99.00
5	100.00	100.00	99.20
6	100.00	100.00	99.90
7	100.00	100.00	100.00
8	100.00	100.00	100.00
9	100.00	100.00	100.00
10	100.00	100.00	100.00
11	100.00	100.00	100.00
12	100.00	100.00	100.00
13	100.00	100.00	100.00
14	100.00	100.00	100.00
15	100.00	100.00	100.00
16	100.00	100.00	100.00
Avg. S_R	100.00	100.00	99.81

rate is obtained by testing the fuzzy system for a thousand noisy data sets. As expected, the success rate improves as the data quality becomes better. Optimizing the fuzzy system for a higher noise level therefore results in good performance at lower noise levels. Furthermore, the fuzzification process of fuzzy logic ensures that the deterioration in the performance of the GFS is gradual with the increase in the noise level. Even though a success rate of 99.81% shown for the noise level for 20% may appear to be very good, it means that in a total of 10,000 samples, 19 cases were not identified correctly. These 19 cases are called false alarms, and one of the main aims of damage detection algorithms is to minimize these false alarms, as they can be expensive and also reduce the faith of the users in the health monitoring system.

To get a physical feel for the change in frequencies due to structural damage and the effect of noise level, we look at the numbers used in rule 4 of Table 3.1 in dimensional form. Rule 4 of Table 3.1 is “slight damage at center.” For this rule $D = 10\%$, i.e., flexural stiffness is reduced from 4×10^8 N-mm^2 for the undamaged beam to 3.6×10^8 N-mm^2 for the damaged beam. For this rule, values of ideal damaged beam frequencies, difference between undamaged beam frequencies and ideal damaged beam frequencies, noisy damaged beam frequencies for positive and negative extreme noise level $\alpha = 0.2$, and difference between ideal damaged beam frequencies and noisy damaged beam frequencies are tabulated in Table 3.3 for eight modes. Here $\omega_1 - \omega_2$ is the change in frequency due to damage. The effect of noise is to allow the noisy damaged frequency to vary between ω_2^n and ω_2^p. For a given

Table 3.3 Effect of adding noise in measurement delta on damaged beam frequency ($d = 8$)

Mode	$\Delta\omega^{\mathrm{fem}}$ (%)	ω_1	ω_2	$\omega_1 - \omega_2$	ω_2^p	ω_2^n	$\omega_2 - \omega_2^p$	$\omega_2 - \omega_2^n$
1	3.72	22.66	21.82	0.84	21.77	21.86	0.04	−0.04
2	3.89	142.03	136.50	5.52	136.22	136.79	0.28	−0.28
3	3.61	397.70	383.34	14.35	382.54	384.13	0.79	−0.79
4	3.65	779.37	750.92	28.44	749.36	752.48	1.55	−1.55
5	3.48	1288.49	1243.65	44.83	1241.07	1246.23	2.57	−2.57
6	4.22	1925.18	1843.93	81.24	1840.08	1847.78	3.85	−3.85
7	4.34	2689.82	2573.08	116.73	2567.70	2578.46	5.37	−5.37
8	3.92	3583.07	3427.56	155.50	3420.40	3434.73	7.16	−7.16

ω_1 = Undamaged beam frequency obtained by FEM (Hz)

ω_2 = Damaged beam frequency obtained by FEM (Hz)

ω_2^p = Damaged beam frequency after adding noise $\alpha = 0.2$, $u = 1$ in $\Delta\omega^{\mathrm{fem}}$ (Hz)

ω_2^n = Damaged beam frequency after adding noise $\alpha = 0.2$, $u = -1$ in $\Delta\omega^{\mathrm{fem}}$ (Hz)

mode, the maximum change in damaged frequency due to noise is given by $\omega_2 - \omega_2^p$ and $\omega_2 - \omega_2^n$. It is clear that, for each mode, the change in damaged frequency due to structural damage is larger than the change due to noise. Of course, this fact is an important requirement for successful damage detection. Even the best soft computing approach cannot perform damage detection if the change in the damage indicator due to damage is less than the noise level. Therefore, the search for good damage indicators is as important in structural health monitoring as the development of soft computing methods.

3.3.2 Test with Different Measurements

The success rate of the fuzzy system is dependent on the number of measurements ($\Delta\omega$'s). Eight frequency measurement deltas were used during the training phase of the GFS. However, for a physical system, the first eight natural frequencies may not be available. Furthermore, some of the higher mode frequencies are difficult to measure accurately, and there is a greater risk of noise contamination in these measurements. We test the automatic rule generating GFS for 4, 5, 6, 7, and 8 measurements. For all these cases, the fuzzy system is optimized using the GA for $\alpha = 0.20$. For $\alpha = 0.20$, the average success rate is 98.68% for four modes and increases to 99.88% for eight modes (Table 3.4). As expected, the addition of measurements increases the accuracy of damage detection. For $\alpha = 0.15$, the average success rate is almost 100% for four modes and for the higher modes. Therefore, for reasonably good quality measurement data,the fuzzy system performs extremely accurately.

The effect of the number of measurements used for damage detection on the average success rate of the fuzzy system is shown in Fig. 3.7. As expected, the

Table 3.4 Success rate for different rules at different numbers of measurements ($\alpha = 0.20$)

Rule No.	S_R $(d = 4)$	S_R $(d = 5)$	S_R $(d = 6)$	S_R $(d = 7)$	S_R $(d = 8)$
1	100.00	100.00	100.00	100.00	100.00
2	100.00	100.00	100.00	100.00	100.00
3	99.00	99.90	99.30	99.90	100.00
4	91.60	87.20	91.50	96.20	99.00
5	90.20	93.70	94.70	98.10	99.20
6	98.40	99.30	99.20	100.00	99.90
7	100.00	100.00	100.00	100.00	100.00
8	100.00	100.00	100.00	100.00	100.00
9	99.80	100.00	100.00	100.00	100.00
10	99.90	100.00	99.90	100.00	100.00
11	100.00	100.00	100.00	100.00	100.00
12	100.00	100.00	100.00	100.00	100.00
13	100.00	100.00	100.00	100.00	100.00
14	100.00	100.00	100.00	100.00	100.00
15	100.00	100.00	100.00	100.00	100.00
16	100.00	100.00	100.00	100.00	100.00
Avg. S_R	98.68	98.76	99.04	99.64	99.88

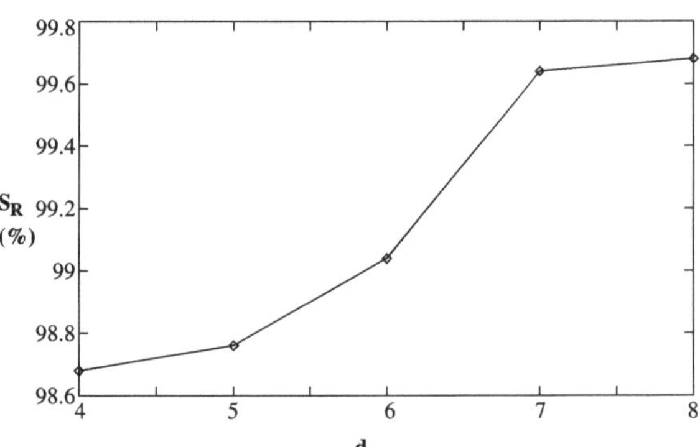

Fig. 3.7 Effect of increase in number of measurement deltas on average success rate ($\alpha = 0.15$)

success rate goes up as more modes are used. Going from four modes to seven modes results in an increase in the success rate of almost 1%.

3.3.3 Missing and Faulty Measurements

An important test for damage detection algorithms is how they perform if one or more measurements are missing or faulty. Consider the case where d' measurement

Table 3.5 Success rate for defective measurement instrument for various d' ($\alpha = 0.15, d = 8$)

Rule No.	S_R ($d' = 1$)	S_R ($d' = 2$)	S_R ($d' = 3$)	S_R ($d' = 4$)
1	100.00	100.00	100.00	100.00
2	100.00	100.00	100.00	99.70
3	100.00	100.00	99.70	99.70
4	100.00	100.00	100.00	99.60
5	100.00	99.80	99.400	98.80
6	100.00	100.00	100.00	99.50
7	100.00	100.00	100.00	100.00
8	100.00	100.00	100.00	100.00
9	100.00	100.00	100.00	100.00
10	100.00	100.00	100.00	99.90
11	100.00	100.00	100.00	100.00
12	100.00	100.00	100.00	100.00
13	100.00	100.00	100.00	100.00
14	100.00	100.00	100.00	100.00
15	100.00	100.00	100.00	100.00
16	100.00	100.00	100.00	100.00
Avg. S_R	100.00	99.99	99.94	99.83

d' = Number of randomly selected measurement deltas which will cross upper or lower limit

deltas exceed a minimum or maximum threshold at a noise level of $\alpha = 0.15$ and eight measurement deltas. For numerical results, we put the minimum threshold at $\Delta\omega = 0$ and the maximum threshold at $\Delta\omega = 15$. These values are selected so that the fuzzy sets defined in Table 3.1 are contained within these thresholds. A thousand samples of noisy data are generated, and for each sample, d' measurement deltas are randomly selected to cross the upper or lower limit for each sample. The success rate for $d' = 1$ is 100% and decreases to 99.83% as d' increases to 4. The results are shown in Table 3.5. This case can represent a scenario where there is missing measurement or bad data such as a bit error. In implementations, such a bad or missing data point(s) may be replaced by a large number (say 100) or a negative number (say -1). Then the fuzzy system would calculate results while neglecting that measurement.

3.3.4 Measurements with High Noise

We consider the case in which noise (α') increases above normal noise at one or two points (randomly selected). As shown in Table 3.6, the average success rate is about 99.92% at $\alpha' = 0.30$ for one randomly selected measurement delta and goes

Table 3.6 Success rate when noise increases at one randomly selected measurement ($\alpha = 0.15$, $d = 8$)

Rule No.	S_R ($\alpha' = 0.30$)	S_R ($\alpha' = 0.40$)	S_R ($\alpha' = 0.50$)	S_R ($\alpha' = 0.60$)
1	100.00	100.00	100.00	100.00
2	100.00	100.00	100.00	99.60
3	100.00	99.70	99.80	98.20
4	99.90	97.50	98.30	95.00
5	99.00	95.40	94.50	89.60
6	99.80	98.60	99.40	96.40
7	100.00	100.00	100.00	100.00
8	100.00	100.00	100.00	100.00
9	100.00	100.00	100.00	100.00
10	100.00	100.00	100.00	100.00
11	100.00	100.00	100.00	100.00
12	100.00	100.00	100.00	100.00
13	100.00	100.00	100.00	100.00
14	100.00	100.00	100.00	100.00
15	100.00	100.00	100.00	100.00
16	100.00	100.00	100.00	100.00
Avg. S_R	99.92	99.45	99.5	98.68

α' = Increased noise level for faulty sensor

on decreasing to 98.68% for $\alpha' = 0.60$. Table 3.7 shows that the average success rate is about 99.51% for $\alpha' = 0.30$ for two randomly selected measurement deltas and goes on decreasing to 96.18% for $\alpha' = 0.60$. These cases can represent a scenario where there are sensor faults or measurement errors effecting one or two modal frequencies. Therefore, it is easier to handle missing measurements and gross errors compared to measurement error. However, the fuzzy system shows a graded degradation in performance. Therefore, it is robust in the presence of significant levels of uncertainty.

3.4 Structural Health Monitoring for Nonuniform Beam

An important advantage of the GFS is that the rule base can easily be obtained for a different structure. To illustrate this, consider a nonuniform beam. Most real physical beam structures are nonuniform in terms of the variation of flexural stiffness and mass per unit length properties. Generally, the root region of a cantilever beam is subject to higher levels of stress. Therefore, designing a cantilever beam which is tapered such that there is more material at the root and less material in the outboard

Table 3.7 Success rate when noise increases at two randomly selected measurements ($\alpha = 0.15$, $d = 8$)

Rule No.	S_R ($\alpha' = 0.30$)	S_R ($\alpha' = 0.40$)	S_R ($\alpha' = 0.50$)	S_R ($\alpha' = 0.60$)
1	100.00	100.00	100.00	100.00
2	99.50	98.40	98.60	96.50
3	99.30	97.70	98.00	93.60
4	97.40	91.90	90.90	80.10
5	96.70	90.50	88.70	79.10
6	99.30	95.40	94.20	89.80
7	100.00	100.00	100.00	100.00
8	100.00	100.00	100.00	100.00
9	100.00	100.00	100.00	99.90
10	100.00	100.00	100.00	99.90
11	100.00	100.00	100.00	100.00
12	100.00	100.00	100.00	100.00
13	100.00	100.00	100.00	100.00
14	100.00	100.00	100.00	100.00
15	100.00	100.00	100.00	100.00
16	100.00	100.00	100.00	100.00
Avg. S_R	99.51	98.37	98.15	96.18

α' = Increased noise level for faulty sensor

regions is common practice. Note, for instance, that the wings of airplanes and the rotor blades of helicopters and wind turbines are tapered beams.

For testing the GFS for a nonuniform beam, the beam illustrated in Fig. 3.5 is tapered by decreasing the area from the fixed end to the free end up to 0.75 times the area at the fixed end. The GA is then used along with the finite element model to create the numerics for the fuzzy sets which maximize the success rate for damage detection of the tapered beam. The inputs are the same eight frequency measurement deltas, and the outputs are the five locations along the beam. Thus, the input and output are exactly the same as those used for the uniform beam earlier in this chapter, but the structure itself is changed to a tapered beam. Automatically generated rules for the tapered beam are shown in Table 3.8. Note that it may have taken a human an enormous amount of time to develop such a rule base. One would have to experiment with thousands of different fuzzy set combinations, and the best success rate obtained using such a process may still not be "optimal." The design of the fuzzy system for structural health monitoring becomes computationally efficient by putting the development of the fuzzy system within the framework of an optimization problem. Table 3.9 shows that the success rate for this case is also excellent. The average success rate for noise level $\alpha = 0.20$ is about 99.06% and increases to 99.92% for a noise level of 0.10. For noise levels of $\alpha = 0.05$ and below, the success rate is 100%.

Table 3.8 Midpoint and standard deviation of fuzzy sets for different rules for tapered beam ($d = 8$, $\alpha = 0.15$)

R.N.	Rule	$\Delta\omega_1$	$\Delta\omega_2$	$\Delta\omega_3$	$\Delta\omega_4$	$\Delta\omega_5$	$\Delta\omega_6$	$\Delta\omega_7$	$\Delta\omega_8$
1	Undamaged	0.00 (0.31)	0.00 (0.33)	0.00 (0.30)	0.00 (0.34)	0.00 (0.26)	0.00 (0.30)	0.00 (0.27)	0.00 (0.31)
2	SLD at Root	4.39 (0.33)	4.26 (0.30)	4.14 (0.34)	3.99 (0.31)	3.45 (0.29)	4.74 (0.27)	4.25 (0.33)	4.15 (0.33)
3	SLD at Inboard	4.07 (0.27)	3.55 (0.29)	3.68 (0.32)	3.83 (0.25)	3.25 (0.26)	4.40 (0.31)	4.01 (0.27)	4.20 (0.34)
4	SLD at Center	3.81 (0.29)	3.87 (0.31)	3.72 (0.27)	3.56 (0.31)	3.40 (0.27)	4.37 (0.32)	4.32 (0.34)	4.05 (0.25)
5	SLD at Outboard	3.65 (0.29)	4.00 (0.31)	3.79 (0.30)	3.52 (0.32)	3.43 (0.27)	4.40 (0.29)	4.25 (0.29)	4.16 (0.30)
6	SLD at Tip	3.59 (0.25)	3.62 (0.32)	3.83 (0.29)	3.98 (0.25)	3.47 (0.30)	4.59 (0.26)	3.97 (0.33)	4.00 (0.26)
7	MOD at Root	9.15 (0.31)	8.90 (0.34)	8.65 (0.26)	8.36 (0.26)	7.32 (0.26)	9.81 (0.25)	8.82 (0.29)	8.64 (0.25)
8	MOD at Inboard	8.52 (0.32)	7.49 (0.32)	7.73 (0.33)	8.03 (0.33)	6.87 (0.28)	9.11 (0.31)	8.32 (0.25)	8.71 (0.27)
9	MOD at Center	8.01 (0.29)	8.11 (0.25)	7.83 (0.28)	7.50 (0.32)	7.15 (0.26)	9.04 (0.34)	8.96 (0.26)	8.43 (0.30)
10	MOD at Outboard	7.70 (0.32)	8.39 (0.29)	7.96 (0.28)	7.44 (0.32)	7.20 (0.25)	9.11 (0.26)	8.84 (0.33)	8.63 (0.25)
11	MOD at Tip	7.59 (0.34)	7.64 (0.32)	8.05 (0.28)	8.33 (0.30)	7.27 (0.32)	9.46 (0.26)	8.25 (0.29)	8.34 (0.31)
12	SVD at Root	14.34 (0.33)	13.96 (0.28)	13.61 (0.32)	13.19 (0.34)	11.68 (0.28)	15.24 (0.31)	13.78 (0.31)	13.52 (0.35)
13	SVD at Inboard	13.43 (0.34)	11.92 (0.29)	12.23 (0.26)	12.66 (0.28)	10.94 (0.32)	14.18 (0.25)	13.02 (0.25)	13.58 (0.29)
14	SVD at Center	12.68 (0.32)	12.81 (0.34)	12.40 (0.32)	11.88 (0.30)	11.32 (0.34)	14.08 (0.30)	13.97 (0.27)	13.23 (0.29)
15	SVD at Outboard	12.22 (0.33)	13.22 (0.28)	12.56 (0.28)	11.83 (0.27)	11.38 (0.32)	14.15 (0.26)	13.85 (0.25)	13.44 (0.32)
16	SVD at Tip	12.06 (0.28)	12.13 (0.34)	12.73 (0.25)	13.10 (0.31)	11.49 (0.29)	14.67 (0.27)	12.91 (0.29)	13.10 (0.26)

SLD = Slight damage

MOD = Moderate damage

SVD = Severe damage

Quantities in parentheses represent standard deviation

Table 3.9 Success rate for different rules at different noise levels for tapered beam ($d = 8$)

Rule No.	$\alpha = 0.10$	$\alpha = 0.15$	$\alpha = 0.20$
1	100.00	100.00	100.00
2	100.00	100.00	100.00
3	100.00	100.00	99.70
4	99.20	92.80	85.50
5	99.50	94.80	87.00
6	100.00	100.00	99.80
7	100.00	100.00	100.00
8	100.00	100.00	100.00
9	100.00	99.90	98.40
10	100.00	100.00	98.60
11	100.00	100.00	100.00
12	100.00	100.00	100.00
13	100.00	100.00	100.00
14	100.00	100.00	100.00
15	100.00	100.00	100.00
16	100.00	100.00	100.00
Avg. S_R	99.92	99.22	98.06

3.4.1 Refined Output Set

The division of a beam into only five parts for structural health monitoring may not be acceptable for many problems. It is often necessary to obtain a higher degree of resolution in damage detection. For instance, for a helicopter rotor blade with a length of 5 m, dividing the beam into five segments gives a damage localization to within 1 m. However, if the beam is divided into ten segments, a localization of 0.5 m is obtained. To obtain more refined results and better accuracy in damage detection we now divide the beam into ten parts. Each of these parts is expressed as a word and given by: ($Root_1$, $Root_2$, $Inboard_1$, $Inboard_2$, $Center_1$, $Center_2$, $Outboard_1$, $Outboard_2$, Tip_1, Tip_2). Then the number of rules will also increase from 16 to 31. The GA is now used to find the optimal fuzzy sets to maximize the success rate for damage detection. The rules obtained from the GFS are shown in Table 3.10. The average success rate is 98.58% for noise level $\alpha = 0.15$. As the number of inputs and outputs grows, fuzzy systems face the problem called "curse of dimensionality," resulting in a great increase in the number of rules. It thus becomes very difficult for a human to design the rule base for large-dimensional fuzzy systems. However, GFSs are very well suited for such problems. Considering the numerics in Table 3.10, it is impossible to imagine that such a complex system could be designed manually by a human.

Table 3.10 Midpoint and standard deviation of fuzzy sets for different rules for refined output set ($d = 8$, $\alpha = 0.15$)

R.N.	Rule	$\Delta\omega_1$	$\Delta\omega_2$	$\Delta\omega_3$	$\Delta\omega_4$	$\Delta\omega_5$	$\Delta\omega_6$	$\Delta\omega_7$	$\Delta\omega_8$	S_R
1	Undamaged	0.00 (0.33)	0.00 (0.35)	0.00 (0.25)	0.00 (0.32)	0.00 (0.27)	0.00 (0.27)	0.00 (0.26)	0.00 (0.25)	100.00
2	SLD at $Root_1$	1.88 (0.32)	1.26 (0.31)	0.87 (0.29)	0.63 (0.32)	0.51 (0.26)	0.49 (0.29)	0.52 (0.30)	0.57 (0.30)	100.00
3	SLD at $Root_2$	1.37 (0.32)	0.22 (0.26)	0.09 (0.34)	0.43 (0.33)	0.71 (0.30)	0.68 (0.25)	0.49 (0.29)	0.39 (0.26)	100.00
4	SLD at $Inboard_1$	0.95 (0.28)	0.06 (0.32)	0.70 (0.32)	0.74 (0.26)	0.25 (0.35)	0.33 (0.30)	0.70 (0.26)	0.63 (0.29)	100.00
5	SLD at $Inboard_2$	0.61 (0.31)	0.48 (0.28)	0.78 (0.26)	0.10 (0.30)	0.71 (0.26)	0.66 (0.27)	0.31 (0.34)	0.63 (0.28)	100.00
6	SLD at $Center_1$	0.35 (0.33)	0.96 (0.32)	0.17 (0.27)	0.74 (0.28)	0.47 (0.25)	0.48 (0.28)	0.63 (0.33)	0.42 (0.27)	87.80
7	SLD at $Center_2$	0.18 (0.26)	1.10 (0.25)	0.22 (0.33)	0.73 (0.32)	0.47 (0.27)	0.48 (0.31)	0.62 (0.34)	0.43 (0.27)	89.50
8	SLD at $Outboard_1$	0.07 (0.33)	0.83 (0.29)	0.98 (0.30)	0.11 (0.32)	0.70 (0.34)	0.65 (0.33)	0.31 (0.30)	0.63 (0.34)	100.00
9	SLD at $Outboard_2$	0.02 (0.27)	0.39 (0.27)	1.10 (0.32)	0.93 (0.31)	0.29 (0.25)	0.32 (0.25)	0.68 (0.25)	0.62 (0.26)	100.00
10	SLD at Tip_1	0.00 (0.29)	0.09 (0.25)	0.42 (0.33)	0.85 (0.25)	1.04 (0.31)	0.90 (0.29)	0.62 (0.32)	0.46 (0.30)	100.00
11	SLD at Tip_2	0.00 (0.26)	0.00 (0.27)	0.02 (0.29)	0.08 (0.30)	0.17 (0.31)	0.29 (0.25)	0.41 (0.26)	0.52 (0.33)	100.00
12	MOD at $Root_1$	4.08 (0.28)	2.66 (0.30)	1.81 (0.30)	1.30 (0.29)	1.08 (0.26)	1.05 (0.34)	1.14 (0.25)	1.25 (0.34)	100.00
13	MOD at $Root_2$	3.01 (0.31)	0.49 (0.28)	0.22 (0.34)	0.94 (0.29)	1.51 (0.25)	1.42 (0.30)	1.03 (0.31)	0.86 (0.31)	100.00
14	MOD at $Inboard_1$	2.10 (0.25)	0.14 (0.26)	1.53 (0.28)	1.59 (0.26)	0.54 (0.27)	0.73 (0.27)	1.47 (0.26)	1.31 (0.34)	100.00
15	MOD at $Inboard_2$	1.36 (0.30)	1.07 (0.29)	1.69 (0.28)	0.23 (0.33)	1.53 (0.29)	1.39 (0.34)	0.71 (0.27)	1.33 (0.25)	100.00
16	MOD at $Center_1$	0.79 (0.28)	2.11 (0.27)	0.37 (0.28)	1.59 (0.25)	1.02 (0.32)	1.05 (0.29)	1.32 (0.30)	0.94 (0.29)	100.00
17	MOD at $Center_2$	0.40 (0.25)	2.42 (0.29)	0.49 (0.30)	1.57 (0.26)	1.01 (0.27)	1.04 (0.25)	1.32 (0.33)	0.94 (0.27)	100.00
18	MOD at $Outboard_1$	0.16 (0.32)	1.84 (0.26)	2.12 (0.33)	0.26 (0.33)	1.50 (0.30)	1.38 (0.25)	0.71 (0.30)	1.32 (0.31)	100.00
19	MOD at $Outboard_2$	0.05 (0.32)	0.88 (0.29)	2.42 (0.31)	1.97 (0.25)	0.64 (0.25)	0.72 (0.32)	1.44 (0.31)	1.29 (0.33)	100.00

Table 3.10 (Continued)

R.N.	Rule	$\Delta\omega_1$	$\Delta\omega_2$	$\Delta\omega_3$	$\Delta\omega_4$	$\Delta\omega_5$	$\Delta\omega_6$	$\Delta\omega_7$	$\Delta\omega_8$	S_R
20	MOD at Tip_1	0.01 (0.25)	0.20 (0.30)	0.95 (0.28)	1.89 (0.33)	2.25 (0.29)	1.89 (0.34)	1.30 (0.29)	1.00 (0.29)	100.00
21	MOD at Tip_2	0.00 (0.27)	0.00 (0.27)	0.05 (0.25)	0.18 (0.28)	0.38 (0.26)	0.68 (0.33)	0.92 (0.33)	1.15 (0.30)	100.00
22	SVD at $Root_1$	6.72 (0.29)	4.22 (0.30)	2.83 (0.26)	2.04 (0.29)	1.72 (0.31)	1.73 (0.34)	1.90 (0.32)	2.07 (0.34)	100.00
23	SVD at $Root_2$	5.01 (0.25)	0.79 (0.28)	0.37 (0.28)	1.57 (0.27)	2.44 (0.25)	2.23 (0.30)	1.64 (0.28)	1.44 (0.31)	100.00
24	SVD at $Inboard_1$	3.52 (0.30)	0.24 (0.27)	2.52 (0.26)	2.55 (0.32)	0.89 (0.32)	1.23 (0.27)	2.34 (0.26)	2.06 (0.28)	100.00
25	SVD at $Inboard_2$	2.29 (0.29)	1.78 (0.35)	2.76 (0.28)	0.40 (0.33)	2.49 (0.33)	2.21 (0.34)	1.21 (0.29)	2.09 (0.31)	100.00
26	SVD at $Center_1$	1.35 (0.25)	3.51 (0.27)	0.63 (0.30)	2.60 (0.25)	1.66 (0.25)	1.70 (0.34)	2.10 (0.33)	1.58 (0.26)	100.00
27	SVD at $Center_2$	0.69 (0.28)	4.03 (0.33)	0.81 (0.27)	2.56 (0.30)	1.64 (0.25)	1.71 (0.26)	2.10 (0.31)	1.58 (0.30)	100.00
28	SVD at $Outboard_1$	0.28 (0.25)	3.10 (0.32)	3.46 (0.32)	0.44 (0.30)	2.43 (0.32)	2.19 (0.30)	1.21 (0.25)	2.09 (0.30)	100.00
29	SVD at $Outboard_2$	0.08 (0.26)	1.50 (0.34)	4.03 (0.25)	3.16 (0.32)	1.05 (0.26)	1.21 (0.30)	2.29 (0.30)	2.03 (0.32)	100.00
30	SVD at Tip_1	0.01 (0.26)	0.34 (0.33)	1.63 (0.28)	3.18 (0.29)	3.65 (0.30)	2.96 (0.28)	2.05 (0.27)	1.66 (0.32)	100.00
31	SVD at Tip_2	0.00 (0.32)	0.01 (0.27)	0.10 (0.26)	0.30 (0.33)	0.66 (0.25)	1.12 (0.30)	1.58 (0.34)	1.94 (0.32)	100.00

SLD = Slight damage

MOD = Moderate damage

SVD = Severe damage

Quantities in parentheses represent standard deviation

Table 3.11 Baseline hingeless blade properties

Number of blades, N_b	4
Radius, R, m	4.94
Hover tip speed, ΩR, m/s	198.12
Mass per unit length, m_o, kg/m	6.46
$EI_y/m_o\Omega^2R^4$	0.0108
$EI_z/m_o\Omega^2R^4$	0.0268
$GJ/m_o\Omega^2R^4$	0.00615
m/m_o	1.0
Lock number	5.2
Solidity, σ	0.07
C_T/σ	0.07
c/R	0.055

3.5 SHM for BO105 Hingeless Helicopter Rotor Blade

Until now, the genetic fuzzy system (GFS) has been developed and tested for a uniform and tapered cantilever beam. To further justify and test the use of this GFS, this system is tested on a realistic example of the blade of a uniform beam model of the BO105 hingeless helicopter rotor. It is assumed that the helicopter is on the ground and the rotor is not rotating. A set of sensors is used to perform a modal analysis of the blade, and the first eight frequencies are extracted. These frequencies are then compared to the saved frequencies of the undamaged blade to find the measurement deltas. These measurement deltas are then fed to the damage detection system. We use finite element simulation to develop the damage detection system. The hingeless rotor blade is fixed at the hub and can be idealized as a cantilever beam. We consider only the out-of-plane or flapwise bending modes of this beam for the damage detection. The properties of this blade are given in Table 3.11. Here, EI_y is the flap stiffness, EI_z is the lag stiffness, and GJ is the torsion stiffness. The Lock number is a non-dimensional ratio of the aerodynamic forces and the inertia forces acting on the rotor. Also, C_T is the non-dimensional thrust coefficient for the helicopter, c is the blade chord, and the rotor solidity is a ratio of the blade area to the area of the rotor disk.

3.5.1 Frequency-Based Damage Detection of Blade

The finite element analysis is performed for the undamaged and all the different sets of damaged blades by seeding the damage at the different locations. The GA is then used to develop the fuzzy system. The rules generated by the GFS are given in Table 3.12. These rules are obtained by optimizing the system for $\alpha = 0.15$ and are tested for $\alpha = 0.1$ to $\alpha = 1.0$. In this problem, we use a high level of noise to test the GFS to its maximum limits. Also, real systems tend to be noisy because of

Table 3.12 Midpoint and standard deviation of fuzzy sets for different rules for BO105 hingeless helicopter rotor blade for frequency-based measurements ($d = 8, \alpha = 0.15$)

R.N.	Rule	$\Delta\omega_1$	$\Delta\omega_2$	$\Delta\omega_3$	$\Delta\omega_4$	$\Delta\omega_5$	$\Delta\omega_6$	$\Delta\omega_7$	$\Delta\omega_8$
1	Undamaged	0.00 (0.32)	0.00 (0.33)	0.00 (0.29)	0.00 (0.34)	0.00 (0.26)	0.00 (0.29)	0.00 (0.27)	0.00 (0.30)
2	SLD at Root	4.45 (0.33)	4.26 (0.30)	4.12 (0.34)	3.98 (0.31)	3.61 (0.29)	4.54 (0.27)	4.21 (0.33)	4.14 (0.33)
3	SLD at Inboard	4.00 (0.27)	3.50 (0.29)	3.71 (0.32)	3.83 (0.25)	3.33 (0.26)	4.19 (0.31)	4.07 (0.27)	4.28 (0.34)
4	SLD at Center	3.72 (0.28)	3.89 (0.27)	3.61 (0.34)	3.64 (0.27)	3.48 (0.32)	4.22 (0.25)	4.34 (0.25)	3.92 (0.31)
5	SLD at Outboard	3.57 (0.33)	3.92 (0.26)	3.84 (0.30)	3.45 (0.34)	3.56 (0.25)	4.32 (0.30)	4.05 (0.30)	4.27 (0.31)
6	SLD at Tip	3.53 (0.25)	3.56 (0.32)	3.74 (0.29)	3.92 (0.25)	3.63 (0.30)	4.49 (0.26)	3.97 (0.33)	3.92 (0.26)
7	MOD at Root	9.26 (0.31)	8.90 (0.34)	8.62 (0.26)	8.35 (0.26)	7.64 (0.25)	9.40 (0.25)	8.74 (0.29)	8.62 (0.25)
8	MOD at Inboard	8.40 (0.32)	7.39 (0.32)	7.78 (0.33)	8.03 (0.33)	7.03 (0.28)	8.69 (0.31)	8.45 (0.25)	8.87 (0.27)
9	MOD at Center	7.83 (0.29)	8.15 (0.25)	7.61 (0.28)	7.65 (0.32)	7.30 (0.26)	8.75 (0.34)	9.00 (0.26)	8.17 (0.30)
10	MOD at Outboard	7.55 (0.32)	8.22 (0.29)	8.04 (0.28)	7.29 (0.32)	7.44 (0.25)	8.96 (0.26)	8.46 (0.33)	8.84 (0.25)
11	MOD at Tip	7.47 (0.34)	7.51 (0.32)	7.86 (0.28)	8.20 (0.30)	7.56 (0.32)	9.30 (0.26)	8.24 (0.29)	8.17 (0.31)
12	SVD at Root	14.49 (0.34)	13.96 (0.28)	13.56 (0.32)	13.17 (0.34)	12.18 (0.28)	14.62 (0.31)	13.67 (0.31)	13.50 (0.35)
13	SVD at Inboard	13.25 (0.34)	11.77 (0.29)	12.30 (0.25)	12.65 (0.28)	11.16 (0.32)	13.58 (0.25)	13.19 (0.25)	13.82 (0.29)
14	SVD at Center	12.42 (0.32)	12.86 (0.34)	12.08 (0.33)	12.08 (0.30)	11.53 (0.34)	13.65 (0.30)	14.05 (0.27)	12.82 (0.29)
15	SVD at Outboard	12.00 (0.33)	12.98 (0.28)	12.67 (0.28)	11.60 (0.27)	11.70 (0.32)	13.97 (0.26)	13.28 (0.25)	13.78 (0.32)
16	SVD at Tip	11.89 (0.28)	11.94 (0.34)	12.45 (0.25)	12.91 (0.31)	11.87 (0.29)	14.49 (0.27)	12.88 (0.29)	12.82 (0.26)

SLD = Slight damage

MOD = Moderate damage

SVD = Severe damage

Quantities in parentheses represent standard deviation

Table 3.13 Success rate for different rules at different noise levels for BO105 hingeless helicopter rotor blade ($d = 8$)

Rule No.	Noise level, α									
	0.10	0.20	0.30	0.40	0.50	0.60	0.70	0.80	0.90	1.00
1	100.00	100.00	100.00	100.00	100.00	100.00	100.00	100.00	100.00	100.00
2	100.00	100.00	100.00	99.50	97.10	94.10	90.70	88.40	84.50	81.40
3	100.00	100.00	96.60	87.10	76.40	66.70	57.60	50.90	44.60	41.70
4	100.00	99.00	88.30	71.20	59.50	51.00	41.40	35.70	30.80	24.70
5	100.00	99.70	95.00	84.80	73.50	68.10	61.60	54.20	51.80	48.10
6	100.00	99.90	94.40	84.50	69.90	64.60	54.60	47.20	42.70	38.50
7	100.00	100.00	100.00	100.00	100.00	100.00	99.50	98.10	96.40	91.50
8	100.00	100.00	100.00	99.90	98.90	96.60	91.50	89.10	81.80	76.40
9	100.00	100.00	99.90	99.80	94.80	90.20	80.50	73.40	68.40	62.00
10	100.00	100.00	100.00	98.80	96.30	90.50	81.10	76.80	68.60	61.40
11	100.00	100.00	100.00	100.00	99.10	96.50	91.40	87.00	82.90	78.30
12	100.00	100.00	100.00	100.00	100.00	100.00	100.00	100.00	100.00	99.90
13	100.00	100.00	100.00	100.00	100.00	99.90	98.30	96.30	92.90	89.10
14	100.00	100.00	100.00	100.00	99.80	99.30	97.80	95.30	92.00	86.80
15	100.00	100.00	100.00	100.00	99.40	98.70	95.40	92.40	86.10	79.10
16	100.00	100.00	100.00	100.00	100.00	99.80	98.70	96.70	93.00	88.70
Avg. S_R	100.00	99.91	98.39	95.41	91.54	88.49	83.76	80.09	76.03	71.73

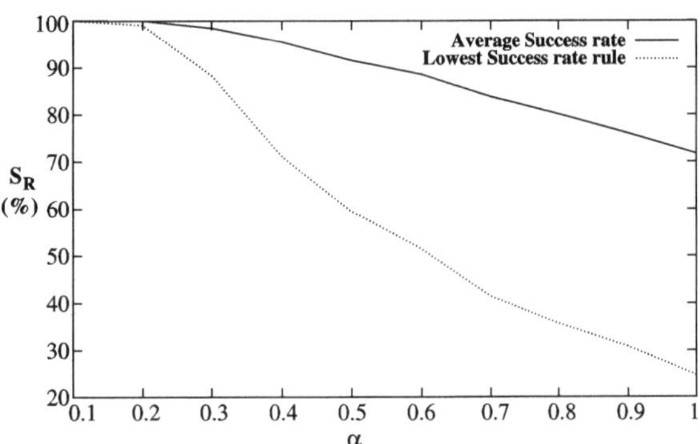

Fig. 3.8 Effect of increasing noise in data on the success rate of GFS ($d = 8$)

both measurement and model uncertainty. The results are shown in Table 3.13. The average success rate is 100% for $\alpha = 0.1$ and 71.73% for $\alpha = 1.0$.

The effect of increasing noise on the average success rate of the fuzzy logic system is shown in Fig. 3.8. It can be observed that the success rate is about 100% until a noise level of $\alpha = 0.10$, after which it starts decreasing. However, even at a noise level of $\alpha = 0.3$, the fuzzy system with eight natural frequencies shows an average success rate of 98.39%. Figure 3.8 also shows the behavior of the lowest success rate case (rule no. 4, slight damage at center). It is observed that at $\alpha = 0.30$, the success rate is 88.30% and at higher noise levels it decreases drastically, even though the average success rate remains high. Thus, for higher α, the system may not work well. The worst diagnosed damage gives the critical noise below which the fuzzy system will work very well. For example, if we want a damage detection accuracy of at least 90%, the critical noise is about 0.3 from Fig. 3.8. It is also observed that for rule no. 1 (for undamaged) and for rule no. 12 to rule no. 16 (for severe damage), the critical noise value will be much higher than that of rule no. 2 to rule no. 6 (for slight damage).

3.5.2 *Mode Shape-Based Damage Detection of Blade*

For the previous results, we have given changes in frequencies as the input data to the GFS. Frequencies are easy to measure and also easy to handle numerically because they are scalars. However, they can be relatively insensitive to low levels of damage in the presence of noise. The mode shapes of a structure represent the "shape" with which the structure vibrates when excited at a given natural frequency. Thus, if the beam is excited at the first natural frequency, it vibrates at the first mode shape. In general, the ith natural frequency of the beam corresponds to the ith mode shape. Some researchers have opined that mode shapes are good indicators of the location of damage. Recall the general eigenvalue problem for structural vibration, where the natural frequencies are the eigenvalues and the mode shapes are the eigenvectors. We now use mode shape data as a measurement delta defined as the percent difference between the norms of the eigenvector Φ:

$$\Delta \|\Phi\| = 100 \frac{\|\Phi_{\text{undamaged}}\| - \|\Phi_{\text{damaged}}\|}{\|\Phi_{\text{undamaged}}\|}. \tag{3.5}$$

Given a computed eigenvector measurement delta $\Delta \|\Phi\|$, random number u in the interval $[-1, 1]$, and a noise level parameter β, the noisy simulated data is given as

$$\Delta \|\Phi\|^{\text{noisy}} = \Delta \|\Phi\|^{\text{fem}} + u\beta. \tag{3.6}$$

The parameter β defines the maximum variance between the computed value of $\Delta \|\Phi\|^{\text{fem}}$ and the simulated measured value $\Delta \|\Phi\|^{\text{noisy}}$, which is a simulation of a practical measurement. Note that since the magnitudes of $\Delta \|\Phi\|$ are much less than those for $\Delta\omega$, the noise level β used here is not directly comparable to α used for frequency-based damage detection.

The rules generated by measurement deltas as norms of the eigenvectors are given in Table 3.14. Some of the measurement deltas are negative because the norm of the eigenvector can increase or decrease from the undamaged level for a stiffness damage, while the eigenvalues will always decrease. The success rates obtained by this method are given in Table 3.15. This method works very well up to a noise level β of 0.10.

Table 3.14 Midpoint and standard deviation of fuzzy sets for different rules for BO105 hingeless helicopter rotor blade by mode shape method ($d = 8, \beta = 0.10$)

R.N.	$\Delta\|\Phi_1\|$	$\Delta\|\Phi_2\|$	$\Delta\|\Phi_3\|$	$\Delta\|\Phi_4\|$	$\Delta\|\Phi_5\|$	$\Delta\|\Phi_6\|$	$\Delta\|\Phi_8\|$	$\Delta\|\Phi_9\|$
1	0.00 (0.26)	0.00 (0.28)	0.00 (0.24)	0.00 (0.29)	0.00 (0.21)	0.00 (0.24)	0.00 (0.22)	0.00 (0.25)
2	−0.28 (0.29)	−0.56 (0.25)	−0.25 (0.29)	−0.66 (0.25)	−0.18 (0.24)	−0.79 (0.23)	0.17 (0.28)	0.97 (0.28)
3	1.22 (0.22)	1.83 (0.24)	0.55 (0.27)	−0.31 (0.20)	0.33 (0.21)	−0.54 (0.26)	0.05 (0.22)	−0.38 (0.28)
4	1.07 (0.28)	1.67 (0.22)	0.51 (0.28)	−0.44 (0.27)	0.35 (0.22)	−0.62 (0.24)	0.53 (0.24)	1.15 (0.23)
5	1.02 (0.22)	1.60 (0.24)	0.47 (0.25)	−0.49 (0.26)	0.33 (0.25)	−0.66 (0.26)	0.52 (0.25)	1.11 (0.22)
6	1.06 (0.28)	1.65 (0.24)	0.52 (0.30)	−0.44 (0.24)	0.42 (0.30)	−0.53 (0.22)	0.76 (0.29)	0.99 (0.25)
7	−0.64 (0.26)	−1.17 (0.29)	−0.56 (0.21)	−1.37 (0.21)	−0.41 (0.21)	−1.68 (0.20)	0.36 (0.24)	1.99 (0.20)
8	2.56 (0.27)	3.84 (0.28)	1.21 (0.28)	−0.69 (0.28)	0.72 (0.23)	−1.18 (0.26)	0.13 (0.20)	−0.85 (0.22)
9	2.22 (0.23)	3.57 (0.25)	1.08 (0.20)	−0.94 (0.27)	0.76 (0.26)	−1.34 (0.26)	1.12 (0.21)	2.38 (0.20)
10	2.09 (0.22)	3.46 (0.29)	1.00 (0.21)	−1.03 (0.20)	0.71 (0.24)	−1.41 (0.23)	1.08 (0.24)	2.32 (0.27)
11	2.22 (0.29)	3.57 (0.27)	1.12 (0.23)	−0.93 (0.25)	0.88 (0.27)	−1.17 (0.21)	1.59 (0.24)	2.05 (0.26)
12	−1.11 (0.29)	−1.81 (0.23)	−0.93 (0.27)	−2.04 (0.29)	−0.70 (0.23)	−2.55 (0.26)	0.56 (0.26)	3.09 (0.30)
13	4.04 (0.29)	6.07 (0.24)	1.98 (0.21)	−1.14 (0.23)	1.19 (0.27)	−1.95 (0.21)	0.25 (0.21)	−1.50 (0.24)
14	3.45 (0.24)	5.76 (0.29)	1.74 (0.27)	−1.46 (0.28)	1.23 (0.21)	−2.15 (0.26)	1.76 (0.20)	3.65 (0.27)
15	3.25 (0.28)	5.63 (0.26)	1.61 (0.26)	−1.57 (0.23)	1.16 (0.23)	−2.22 (0.22)	1.72 (0.21)	3.59 (0.27)
16	3.50 (0.23)	5.78 (0.29)	1.82 (0.21)	−1.47 (0.26)	1.40 (0.24)	−1.96 (0.22)	2.50 (0.24)	3.11 (0.21)

Rules are defined similarly as in Table 3.1

Quantities in parentheses represent standard deviation

Table 3.15 Success rate for different rules at different noise levels for BO105 hingeless helicopter rotor blade by mode shape method ($d = 8$)

Rule No.	$\beta = 0.00$	$\beta = 0.05$	$\beta = 0.10$	$\beta = 0.15$
1	100.00	100.00	100.00	100.00
2	100.00	100.00	100.00	100.00
3	100.00	100.00	100.00	100.00
4	100.00	99.00	87.80	73.20
5	100.00	99.10	85.30	70.90
6	100.00	100.00	99.90	96.50
7	100.00	100.00	100.00	100.00
8	100.00	100.00	100.00	100.00
9	100.00	100.00	96.90	88.30
10	100.00	100.00	98.10	90.40
11	100.00	100.00	100.00	100.00
12	100.00	100.00	100.00	100.00
13	100.00	100.00	100.00	100.00
14	100.00	100.00	99.80	97.70
15	100.00	100.00	99.80	94.20
16	100.00	100.00	100.00	100.00
Avg. S_R	100.00	99.88	97.97	94.45

3.6 Summary

This chapter illustrates the use of the genetic fuzzy system for damage detection in a beam, which is one of the simplest and most important structures. First, the structural health monitoring system is developed and tested for a uniform beam using the change in natural frequencies relative to the undamaged structure as the damage indicators. Various conditions involving noisy data, different measurement sets, different levels of damage resolution, and faulty measurements are used, and the genetic fuzzy system is found to perform well. The structural health monitoring system is then developed for a tapered beam and a BO105 hingeless helicopter rotor blade. For each structure, a genetic algorithm is used to develop the fuzzy system to maximize the success rate of damage detection. The complexity of the fuzzy rules clearly highlights the need for automation. Finally, mode shapes are used in place of frequencies to perform damage detection. The genetic fuzzy system performs well for all the cases considered in this chapter and shows a gradual decrease in performance as noise increases. The genetic fuzzy system detects damage with an accuracy of nearly 100% for low to moderate levels of noise in the data. The problems in the chapter focus on a beam structure made of isotropic material. Both the damage model and the mathematical model of the structure are kept simple to focus on the genetic fuzzy system development process. In the next chapter, a more complex structure (thin-walled tubular beam) made of composite material will be considered to illustrate damage detection using the genetic fuzzy system.

References

1. Chandrupatla, T.R., Belegundu, A.D.: Introduction to Finite Elements in Engineering. Prentice-Hall, Englewood Cliffs (2001)
2. Reddy, J.N.: An Introduction to the Finite Element Method. McGraw-Hill Higher Education, New York (2006)
3. Bathe, K.J.: Finite Element Procedures. Prentice-Hall, Englewood Cliffs (1996)
4. Zienkiewicz, O.C., Tylor, R.L.: The Finite Element Method. Butterworth-Heinemann, Oxford (2000)
5. Ganguli, R.: Health monitoring of helicopter rotor in forward flight using fuzzy logic. AIAA J. **40**(12), 2773–2781 (2002)
6. Ganguli, R.: A fuzzy logic system for ground based structural health monitoring of a helicopter rotor using modal data. J. Intell. Mater. Syst. Struct. **12**(6), 397–408 (2001)
7. Pawar, P.P., Ganguli, R.: Genetic fuzzy system for damage detection in beams and helicopter rotor blades. Comput. Methods Appl. Mech. Eng. **192**(16–18), 2031–2057 (2003)
8. Chandrasekhar, M., Ganguli, R.: Uncertainty handling in structural damage detection using fuzzy logic and probabilistic simulation. Mech. Syst. Signal Process. **23**(2), 384–404 (2009)

Chapter 4
Structural Health Monitoring of Composite Tubes

This chapter investigates the effects of structural damage on the behavior of thin-walled composite tubes. A tube of hollow circular cross section made of composite material is considered. Composites are materials formed by combining at least two different types of materials so as to take advantage of some properties of each of them. For example, fiber reinforced polymer composites are made by reinforcing a polymer matrix material with fibers. The polymer matrix may be epoxy, and the fibers could be carbon in the form of graphite. Here, the fibers contribute to the stiffness and the polymer to the volume and lightweight properties. Composite materials have become popular because of their low weight, superior fatigue properties, damage tolerance, and crashworthiness. Their initial use was in the aerospace industry, but now composites are being used in many fields of structural engineering. Despite these advantages, composites also create complications due to their more complex modeling requirements as compared to metals. Unlike metals, the material properties of composites are directional in nature. Composites are susceptible to several different damage mechanisms. The three most important damage mechanisms which occur in composites are matrix cracking, delamination, and fiber breakage. Generally, the first damage which occurs in composite structures is matrix cracking. As matrix cracks build up within the structure, they cause a degradation in the stiffness. However, after a critical matrix crack saturation density is reached, the accumulation of more matrix cracks causes negligible change in the structural stiffness. This point, called matrix crack saturation, is an important point which shows that more serious damage mechanisms are likely to begin. Typically, matrix crack saturation is followed by delamination/debonding and fiber breakage. Therefore, the values of damage indicators at the point of matrix crack saturation can be used as a threshold to put the structure under closer watch. In this chapter, matrix cracking in the composite tube is considered as the damage. Since both the structure and damage models are more complicated than the isotropic beams considered in the previous chapter, the mathematical model development is more elaborate in this chapter. Thus, this chapter illustrates the development of the genetic fuzzy system for a more complicated structure compared to the previous chapter.

P.M. Pawar, R. Ganguli, *Structural Health Monitoring Using Genetic Fuzzy Systems*,
DOI 10.1007/978-0-85729-907-9_4,

4.1 Matrix Cracking in Hollow Circular Cross Section

We need to develop a mathematical model which predicts the change in modal properties such as frequencies of a composite tube due to localized matrix cracks. This is achieved by including a matrix crack model in the structural model of the composite tube. The composite tube has a hollow circular cross section with uniform diameter and thickness along the length. Such composite tubes find many industrial applications. For example, they can be used as electrical transmission poles [1, 2] and as parts of wind turbine towers [3]. Circular composite tubes are also being used for crash absorption components in trains [4]. There is a trend to replace metal tubes with composite tubes in applications due to their lower weight and better crashworthiness properties. Thus, composite tubes are important structures. However, damage growth in these tubes can result in undesirable stiffness loss and resulting changes in displacement and natural frequencies which may also be undesirable. The cross section of the tube is shown in Fig. 4.1. The dynamic analysis of the composite tube is done by using a beam finite element analysis discussed in the previous chapter for the Euler–Bernoulli cantilever beam (Sect. 3.1). The tube is long and slender; thus, an Euler–Bernoulli beam model is appropriate. The effective elastic modulus required for beam analysis with matrix crack damage is discussed next.

4.1.1 Effective Elastic Modulus

This section derives an effective elastic modulus of a thin-walled composite structure which can be useful for simple thin-walled composite beams such as a hollow circular cross section with the wall having the $[\pm\theta_m/90_n]_s$ family of composite laminates. For the one-dimensional (1D) model, the effective elastic modulus is required which accounts for all composite properties including the effect of matrix cracking. The effective elastic modulus is formulated considering the small cross section of the beam as a 2D composite plate.

Modeling of composite structures is done using the classical laminated plate theory. A composite laminate is made by stacking together many laminas. Each lamina or ply has an associated ply angle at which the fibers are oriented. A key advantage of composites is that the fibers can be tailored for improved properties of the structure. Details about composite materials can be obtained from books such as the one by Jones [5]. Using the classical laminate theory, the coefficients of a generally orthotropic plate can be determined. The constitutive equation for the laminate is

$$\begin{Bmatrix} \mathbf{N} \\ \mathbf{M} \end{Bmatrix} = \begin{bmatrix} \mathbf{A} & \mathbf{B} \\ \mathbf{B} & \mathbf{D} \end{bmatrix} \begin{Bmatrix} \boldsymbol{\varepsilon} \\ \mathbf{k} \end{Bmatrix}. \tag{4.1}$$

Here $\mathbf{N}$ and $\mathbf{M}$ are vectors representing three forces and three moments, respectively. Also, $\boldsymbol{\varepsilon}$ and $\mathbf{k}$ are vectors of strain and curvature, respectively. The extension bending stiffness matrix (**B**) is a null matrix for a symmetrical laminate. From the equideformability hypothesis [6] of the laminate, one can conclude that

$$\begin{Bmatrix} N_x \\ N_y \\ N_{xy} \end{Bmatrix} = \begin{bmatrix} A_{11} & A_{12} & A_{16} \\ A_{12} & A_{22} & A_{26} \\ A_{16} & A_{26} & A_{66} \end{bmatrix} \begin{Bmatrix} \varepsilon_x \\ \varepsilon_y \\ \varepsilon_{xy} \end{Bmatrix} \tag{4.2}$$

Inverting (4.2) we obtain

$$\begin{Bmatrix} \varepsilon_x \\ \varepsilon_y \\ \varepsilon_{xy} \end{Bmatrix} = \begin{bmatrix} a'_{11} & a'_{12} & a'_{16} \\ a'_{12} & a'_{22} & a'_{26} \\ a'_{16} & a'_{26} & a'_{66} \end{bmatrix} \begin{Bmatrix} N_x \\ N_y \\ N_{xy} \end{Bmatrix} \tag{4.3}$$

Note that the shear-extension coupling terms A_{16} and A_{26} in (4.2) vanish for the $[\pm\theta_m/90_n]_s$ family of composites, and the term a'_{11} in (4.3) includes the shear effect due to layer stiffness transformation. The terms a'_{16} and a'_{26} are also zero in this case. The effective longitudinal modulus E_{eff} [7] can then be calculated from (4.3) as

$$E_{\text{eff}} = \frac{1}{a'_{11}\, t}. \tag{4.4}$$

For structures having a wall thickness smaller than the beam dimension, the rigidity of the beam is minimally dependent on the local rigidity of the wall. This fact has been used to simplify the modeling by several researchers. Estivalezes and Barrau [8] mentioned that for the thin-walled cross section with respect to other dimensions of the beam, the local rigidities can be neglected; i.e., the thin-walled beam can be analyzed by using only the extensional stiffness matrix. Recently, Ferrero [6] used the constitutive relation based on the extensional stiffness matrix on the basis of the equideformability hypothesis of the laminate. Hence, the effective modulus E_{eff} can be used for axial as well as flexural problems without loss of material characteristics that affect the structural behavior of the member.

4.1.2 Matrix Crack Model ($[\pm\theta_m/90_n]$ Family of Composites)

The damage model used in this chapter is a matrix crack model for cross-ply laminates. This model is developed based on the Nuismer and Tan [9] study for matrix cracking in the $[\pm\theta_m/90_n]$ family of composites. The identification of an appropriate damage model is a crucial step in the development of the model-based structural health monitoring (SHM) system. The model should be as simple as possible, but should still capture the basic physics of the problem accurately. Typically, a thorough search of the literature should be conducted and the appropriate damage model should be selected. In some cases no usable model may be available, and a damage model may need to be developed for SHM. In many cases, damage models developed to model the intricate physics of the problem using micromechanics, atomistic, and multiscale approaches may be too computationally intensive to be used for model-based SHM.

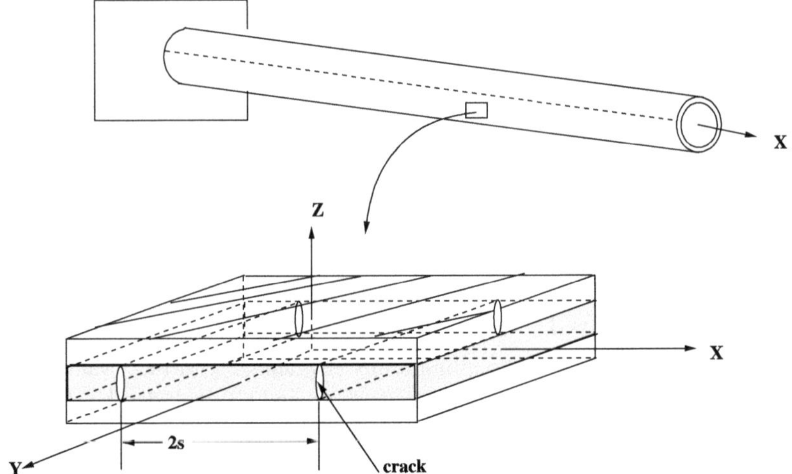

Fig. 4.1 A hollow circular cantilever beam showing small cross section of wall, composite laminate containing a cracked ply

Fig. 4.2 One-quarter of the composite laminate unit cell

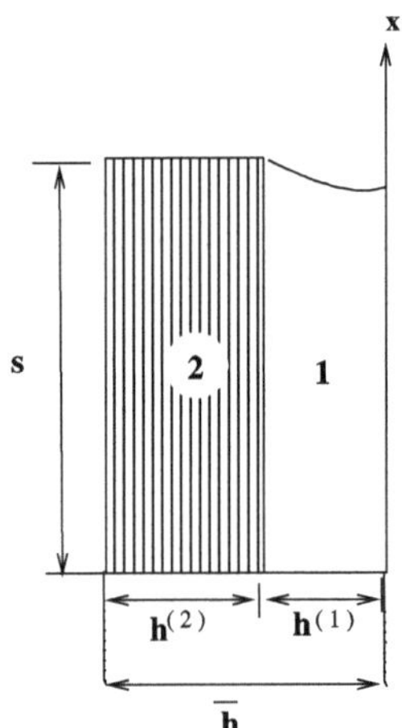

To understand the damage modeling, consider a composite beam subjected to a general in-plane loading $\bar{\sigma}_x$, $\bar{\sigma}_y$, $\bar{\tau}_{xy}$ and containing a cracked central lamina, as shown in Fig. 4.1. The center layer consists of a 90°-ply group and top and bottom layers of a $\pm\theta°$-ply group. Matrix cracks are assumed to exist in the 90°-ply group with a uniform crack spacing of 2 s. Further, for symmetry reasons, only one quarter of the repeating interval of the laminate is modeled. The modeled portion of length s is divided into two sub laminates, as shown in Fig. 4.2. In this figure, sub laminate 1 denotes the 90°-ply group and sub-laminate 2 denotes the $\pm\theta°$-ply group.

The key to this approach lies in defining average properties across the lamina thickness:

$$(\bar{X})^{(i)} = \frac{1}{h^{(i)}} \int_{h^{(i)}} (X)^{(i)}\, dz, \tag{4.5}$$

where a bar $(\bar{X})$ is used to designate the through-the-thickness average of a particular variable X and the superscript (i) denotes sub-laminate 1 or 2.

One more assumption is that there is a simple linear variation in the out-of-plane shear stresses across the lamina thickness. Careful investigation of the governing equations in Nuismer and Tan [9] reveals that the solution for the in-plane normal response of the laminate decouples from the solution for the in-plane shear response due to the assumed orthotropy of the laminas. Therefore, each response is considered separately.

The reduced laminate stiffness matrix, having the effect of matrix cracking in 90° plies, is given as

$$\bar{A} = \begin{bmatrix} \bar{A}_{11} & \bar{A}_{12} & 0 \\ \bar{A}_{12} & \bar{A}_{22} & 0 \\ 0 & 0 & \bar{A}_{66} \end{bmatrix}, \tag{4.6}$$

where the terms of the above matrix can be written as

$$\bar{A}_{11} = \frac{h^{(1)}\bar{Q}_{11}^{(1)} + h^{(2)}\bar{Q}_{11}^{(2)}}{\beta_2 \bar{h}}, \tag{4.7}$$

$$\bar{A}_{12} = \frac{\beta_1 h^{(1)}\bar{Q}_{12}^{(1)} + \beta_2 h^{(2)}\bar{Q}_{12}^{(2)}}{\beta_2 \bar{h}}, \tag{4.8}$$

$$\bar{A}_{22} = \frac{h^{(1)}\bar{Q}_{22}^{(1)} + h^{(2)}\bar{Q}_{22}^{(2)}}{\bar{h}} - \left(\frac{\beta_2 - \beta_1}{\beta_2}\right)\frac{h^{(1)}}{\bar{h}}\frac{\bar{Q}_{12}^{(1)\,2}}{\bar{Q}_{11}^{(1)}}, \tag{4.9}$$

$$\bar{A}_{66} = \frac{h^{(1)}\bar{Q}_{66}^{(1)} + h^{(2)}\bar{Q}_{66}^{(2)}}{\beta_4 \bar{h}}, \tag{4.10}$$

and the values of β_1, β_2, and β_4 are

$$\beta_1 = 1 - \frac{\tanh \alpha_1 s}{\alpha_1 s}, \tag{4.11}$$

$$\beta_2 = 1 + \frac{h^{(1)}Q_{11}^{(1)}}{h^{(2)}Q_{11}^{(2)}} \cdot \frac{\tanh \alpha_1 s}{\alpha_1 s}, \tag{4.12}$$

$$\beta_4 = 1 + \frac{h^{(1)}Q_{66}^{(1)}}{h^{(2)}Q_{66}^{(2)}} \cdot \frac{\tanh \alpha_2 s}{\alpha_2 s}, \tag{4.13}$$

and where the values of α_1 and α_2 are

$$\alpha_1^2 = \frac{3Q_{55}^{(1)}Q_{55}^{(2)}}{h^{(1)}Q_{55}^{(2)} + h^{(2)}Q_{55}^{(1)}} \left(\frac{h^{(1)}Q_{11}^{(1)} + h^{(2)}Q_{11}^{(2)}}{h^{(1)}h^{(2)}Q_{11}^{(1)}Q_{11}^{(2)}}\right), \tag{4.14}$$

$$\alpha_2^2 = \frac{3Q_{44}^{(1)}Q_{44}^{(2)}}{h^{(1)}Q_{44}^{(2)} + h^{(2)}Q_{44}^{(1)}} \left(\frac{h^{(1)}Q_{66}^{(1)} + h^{(2)}Q_{66}^{(2)}}{h^{(1)}h^{(2)}Q_{66}^{(1)}Q_{66}^{(2)}} \right). \tag{4.15}$$

$\bar{\mathbf{A}}$ is obtained by averaging both the sub-laminates to account for the effect of matrix cracking. Therefore, $\bar{\mathbf{A}}$ is the reduced stiffness matrix for the whole laminate. Hence, the extensional stiffness matrix, having the effect of matrix cracking, can be obtained directly as

$$\mathbf{A} = t\bar{\mathbf{A}}. \tag{4.16}$$

This matrix crack model is valid only for the $[\pm\theta_m/90_n]s$ family of composites with matrix cracking in 90° laminates. The model is simple and easy to apply for preliminary investigations on the effect of matrix cracking on composite structure stiffness. The model is also limited to composite structure theories using the **A** matrix. Some thin-walled beam theories for aerospace structures use the **A** matrix only, and the model discussed above is applicable to such problems. Thus, we now have a model of the undamaged composite tube and a way of simulating matrix cracks in the composite tube. Numerical experiments for the undamaged and damaged composite tube can now be performed, and these experiments will allow us to perform model-based diagnostics.

A sample MATLAB code for calculating the effective Young's modulus for a given matrix crack density (refer to (4.4)) *is given in file "matrix_cracking_code.m".*

4.2 Modal Analysis

Computational modal analysis is performed for the composite tube. The tube length is 18 m and the outer diameter of the circular cross section is 600 mm. These properties are taken from Polyzois et al. [7], and this structure was selected since results to validate the analysis of the undamaged composite tube (without matrix cracks) are also given by Polyzois et al. [7]. Furthermore, the structure is a realistic physical structure representing a long composite pole of the kind used for electrical transmission wires. Material properties corresponding to glass epoxy are $E_1 = 48$ Gpa, $E_2 = 13.30$ Gpa, $\nu_{12} = 0.235$, and $G_{12} = 5.17$ Gpa, where 1 denotes the fiber direction and 2 denotes the transverse direction. The material density $\rho = 1.94\ \mathrm{gm/cm^3}$, and the wall contains 50 plies, where each ply is 0.22 mm thick. The total thickness of the beam wall cross section is 11 mm. We can see that the thickness of the beam wall of 11 mm is much less than the diameter of the cross section of 600 mm. This is a typical thin-walled structure which makes very good use of the material for giving desired stiffness and strength properties.

The baseline undamaged finite element model converges at eight number of elements for the undamaged structure. The natural frequencies of the fundamental mode are compared with the frequencies of the Polyzois et al. [7] model of composite beam with hollow circular cross section computed for various θ values of $[\pm\theta]_{25}$ composite (see Table 4.1). Polyzois et al. [7] validated this beam model by comparing the natural frequency of the fundamental mode obtained by 3D shell

Table 4.1 For the typical $[\pm\theta_m]_{25}$ family of composites, comparison of fundamental frequencies (rad/sec) for undamaged structure

$\pm\theta$	0°	5°	10°	15°	20°	25°	30°	35°	40°	45°
Present ω_p	11.45	11.37	11.12	10.71	10.14	9.42	8.62	7.83	7.12	6.58
Polyzois et al. ω_l	11.35	11.28	11.04	10.63	10.06	9.35	8.55	7.76	7.07	6.53
Comparison $\frac{(\omega_p-\omega_l)\times 100}{\omega_l}$ %	0.88	0.80	0.72	0.75	0.79	0.75	0.82	0.90	0.71	0.77

Table 4.2 For a typical $[\pm\theta_m/90_n]_s$ family of composites, effective modulus (GPa) and fundamental frequencies (rad/sec) for undamaged structure

n	θ					
	0°		30°		60°	
	E_{eff}	ω	E_{eff}	ω	E_{eff}	ω
5	41.208	10.610	26.161	8.453	13.306	6.029
9	35.675	9.872	24.210	8.132	13.509	6.075
17	24.514	8.183	19.118	7.226	13.574	6.089

element analysis of the beam using the ANSYS finite element program. Our predictions match those of Polyzois et al. [7] within 0.71–0.9%. Therefore, the baseline model for the undamaged composite pole is validated.

Results are first obtained for a uniform pole with matrix cracks. The effects of matrix cracking in the effective matrix are included by substituting the matrix **A** from (4.16). For the typical $[\pm\theta_m/90_n]_s$ family of composites, the effective elastic modulus and fundamental frequencies for the undamaged structure at $\theta = 0$, 30, 60 and $n = 5$, 9, 17 are shown in Table 4.2. The stiffness decreases with an increase in constraining angle θ and the number of 90° plies (n). Figure 4.3 shows the variation in effective modulus and frequency delta (change between damaged and undamaged frequency) due to an increase in crack density for several θ and n combinations.

The crack density is varied from 0–50 cracks/10 cm. From these plots it is observed that the variation of fundamental frequencies is dependent on the crack density, the number of 90° plies, and also the angle of the constraining layer (θ). Using these plots, the range of crack density for each damage level is decided. This type of parametric study done for a structure gives a clear idea about how the damage indicator is related to the damage. In the absence of experimental data, the trends obtained by the parametric study can be used to evaluate the mathematical model using physical reasoning. For the given huge composite transmission pole, performing experiments on the structure would be difficult and expensive.

Next, results are obtained for a nonuniform beam with ten segments, each of length 10% of the beam length. The nonuniformity is in the stiffness along the beam length. Each segment contains two finite elements and the tube is modeled using 20 finite elements. Matrix crack is assumed to accumulate individually in each segment; i.e., we are considering localized matrix cracking. By considering

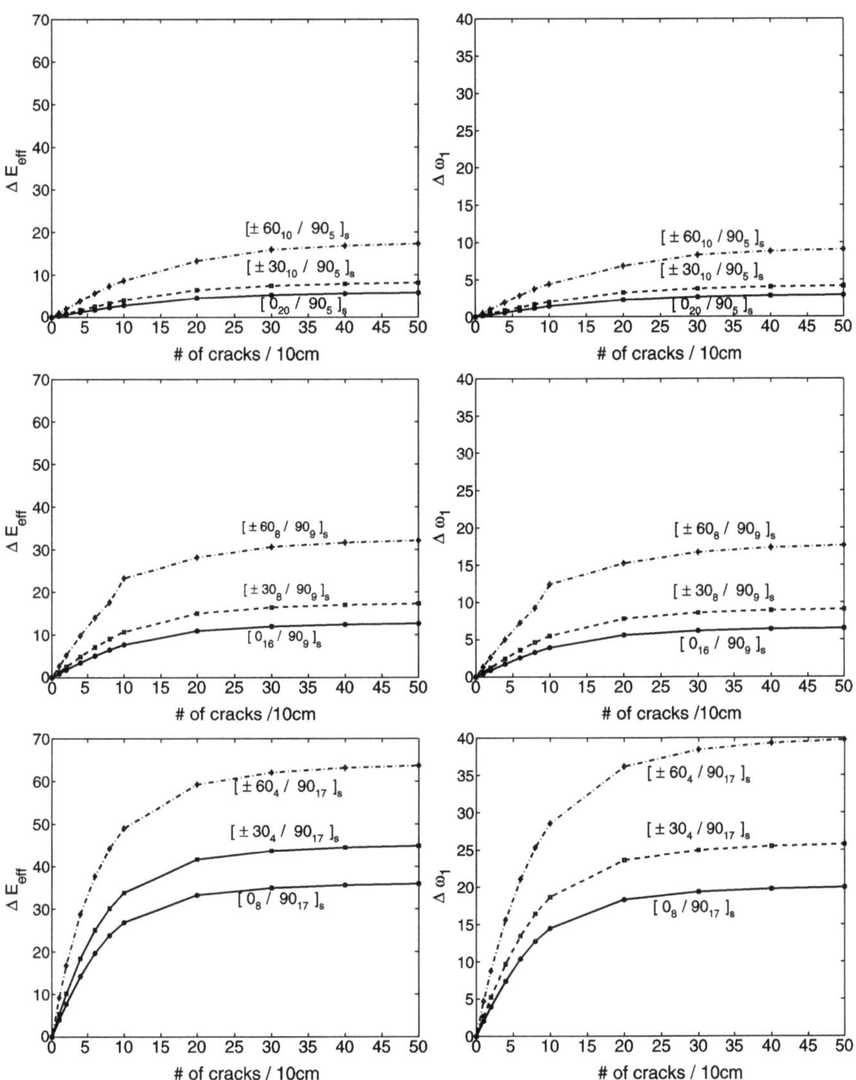

Fig. 4.3 Effect of number of cracks on percentage change in effective modulus (ΔE_{eff}) and percentage change in fundamental frequency ($\Delta\omega$) for $[\pm\theta_m/90_n]_s$ composites at $\theta = 0$, 30, 60, and $n = 5$, 9, 17

damage in each segment, the change in frequency for the first eight modes as $\Delta\omega_1, \Delta\omega_2, \ldots, \Delta\omega_8$ is obtained. In this way, the measurement deltas ($\Delta\omega^{\text{fem}}$) are obtained for 31 rules of the genetic fuzzy system. The effect on the first four modes of frequencies of various crack densities and at various matrix cracking locations for $[0_{16}/90_9]_s$, $[\pm 30_8/90_9]_s$, and $[\pm 60_8/90_9]_s$ composite structures are shown in surface plots of variation of change in frequency ($\Delta\omega$) with respect to damage location and crack density (see Fig. 4.4).

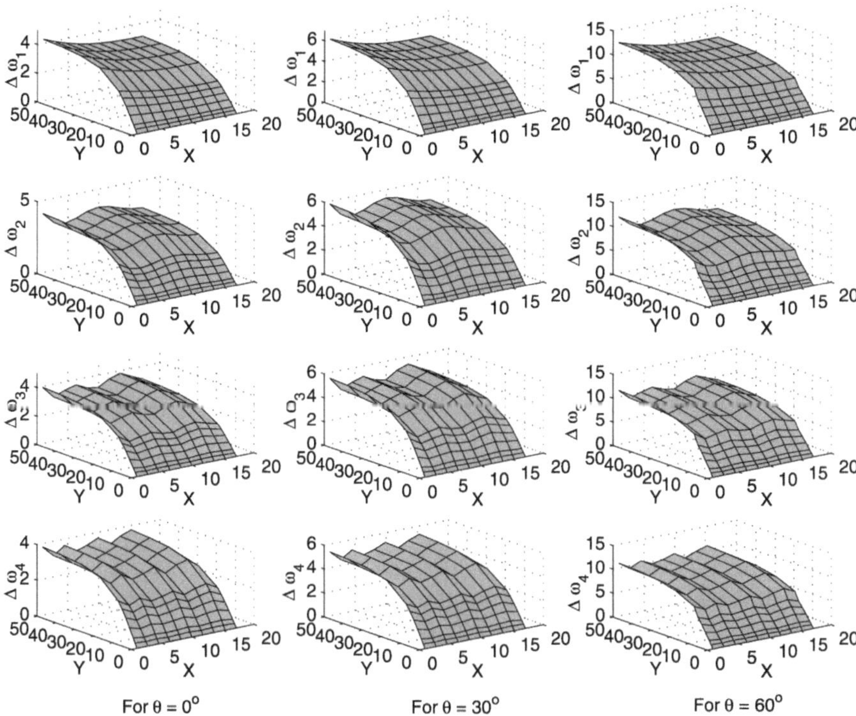

Fig. 4.4 Effect of number of cracks at various locations on percentage change in frequencies for $[\pm\theta_8/90_9]_s$ composite (number of cracks/10 cm along X-direction and crack location (m) along Y direction)

A sample MATLAB code for calculating change in frequencies at different crack densities is given in file "frequency_calculation_code.m", for which the input file is "model2.inp" and the output file is "model2.out".

4.3 Damage Detection in Composite Structure

In the previous chapter, the genetic fuzzy system was demonstrated for damage detection in beam-type structures. However, the damage was modeled as a decrease in stiffness at the damage location, which is typically done for isotropic structures. In this section, a genetic fuzzy system is developed for the thin-walled composite tube by directly relating the internal state of the composite structure in the form of matrix crack density to the measurement deltas.

4.3.1 Development of Genetic Fuzzy System

In this section, we formulate the genetic fuzzy system (GFS) for the detection of crack density and location of cracks for a composite tube. Inputs to the GFS are measurement deltas; outputs are matrix crack density and locations of cracks. The objective is to find the mapping between the measurement deltas and the ten locations of matrix cracking. Numerical simulation of the measurement deltas (change in frequencies) for the composite tube is explained in Sect. 4.2.

In this fuzzy system, the matrix crack location is expressed in words. For example, the location of matrix cracking "L_{10}" ranges from 0% to 10% of the beam from the fixed end, "L_{20}" from 10% to 20%, and so on up to "L_{100}" from 90% to 100%. There are ten damage locations along the tube. Recall that the tube is 18 m long, meaning that each damage location is 1.8 m in length. Matrix cracks are often initiated by low velocity impact [10], which can occur on such composite tubes. To get a degree of resolution of the extent of matrix cracking, each of these matrix cracking locations is allowed several levels of damage and is split into linguistic variables. These classifications are based on the numerical results obtained for matrix cracking. From the plots of change in frequency and matrix crack density shown in Fig. 4.4, the matrix crack density can be broadly classified as "Undamaged" for zero crack density, "Slight damage" for 1–5 crack density, "Moderate damage" for 5–10 crack density, "Severe damage" for 10–20 crack density, and "Very severe damage" for crack density more than 20. Crack density in this chapter refers to the number of cracks per 10 cm. Note that the crack densities selected here are for the composite tube and will change for different structures.

The measurement deltas $\Delta\omega_1, \Delta\omega_2, \ldots, \Delta\omega_8$ are also treated as fuzzy variables. These are the changes in the first eight natural frequencies of the structure relative to the undamaged baseline condition. Fuzzy sets with Gaussian membership functions are used to define these input variables. A change in frequency (measurement delta) is calculated by finite element simulation for a combination of ten different locations and four different levels of damage (Undamaged, Slight Damage, Moderate Damage, and Severe Damage). Crack densities greater than 20 represent very severe damage and are excluded.

Rules for the fuzzy system are obtained by fuzzification of the numerical values obtained from finite element analysis. The fuzzy sets corresponding to $\Delta\omega_1, \Delta\omega_2, \ldots, \Delta\omega_8$ are generated by taking the $\Delta\omega$'s obtained by the finite element solution as midpoints of the membership function corresponding to a location of matrix cracking and damage level. This strategy for selecting the midpoint ensures that the maximum degree of membership ($\mu = 1$) for each fuzzy set occurs at the values of $\Delta\omega$ since the Gaussian function is highest at the midpoint. The standard deviation of the Gaussian membership functions are calculated using a genetic algorithm for maximization of the success rate.

In this way, the "Undamaged" level of damage is represented by one rule and the three damage levels "Slight," "Moderate," and "Severe" at ten locations are represented by a total of ten rules each. Therefore, the complete matrix crack detection system can be represented by 31 rules. For example "Slight" damage at the 0–10% part of the beam can be written as "Slight L_{10}." Rules of the GFS along with midpoints and standard deviations of its Gaussian membership functions for $[0_{16}/90_9]_s$ composite are given in Table 4.3. Rules for all other cases are formulated in a similar manner by solving the optimization problem. Once the GFS is developed, it can

Table 4.3 Midpoints and standard deviations for different rules for typical $[0_{16}/90_9]_s$ family of composites

R.N.	$\Delta\omega_1$	$\Delta\omega_2$	$\Delta\omega_3$	$\Delta\omega_4$	$\Delta\omega_5$	$\Delta\omega_6$	$\Delta\omega_7$	$\Delta\omega_8$
1	0.00 (0.33)	0.00 (0.34)	0.00 (0.32)	0.00 (0.35)	0.00 (0.31)	0.00 (0.32)	0.00 (0.31)	0.00 (0.33)
2	1.11 (0.34)	1.06 (0.33)	1.02 (0.34)	0.99 (0.33)	0.96 (0.32)	0.94 (0.31)	0.92 (0.34)	0.92 (0.34)
3	1.03 (0.30)	0.85 (0.32)	0.80 (0.33)	0.82 (0.34)	0.88 (0.30)	0.93 (0.33)	0.94 (0.34)	0.92 (0.34)
4	0.96(0.32)	0.79 (0.30)	0.88 (0.31)	0.94 (0.31)	0.87 (0.32)	0.81 (0.31)	0.88 (0.32)	0.96 (0.31)
5	0.90 (0.34)	0.85 (0.34)	0.94 (0.33)	0.82 (0.34)	0.87 (0.32)	0.96 (0.35)	0.85 (0.34)	0.87 (0.31)
6	0.86 (0.30)	0.93 (0.30)	0.84 (0.34)	0.88 (0.33)	0.93 (0.31)	0.82 (0.30)	0.96 (0.33)	0.84 (0.30)
7	0.82 (0.31)	0.97 (0.32)	0.80 (0.31)	0.95 (0.30)	0.82 (0.31)	0.94 (0.31)	0.86 (0.33)	0.93 (0.34)
8	0.81 (0.31)	0.94 (0.34)	0.92 (0.34)	0.81 (0.34)	0.96 (0.32)	0.86 (0.33)	0.86 (0.30)	0.96 (0.31)
9	0.79 (0.31)	0.87 (0.32)	0.98 (0.30)	0.91 (0.32)	0.81 (0.31)	0.89 (0.33)	0.96 (0.33)	0.87(0.31)
10	0.79 (0.33)	0.81 (0.30)	0.89 (0.33)	0.97 (0.33)	0.98 (0.34)	0.91 (0.31)	0.84 (0.32)	0.84 (0.32)
11	0.79 (0.33)	0.79 (0.31)	0.80 (0.30)	0.82 (0.32)	0.85 (0.31)	0.89 (0.34)	0.93 (0.32)	0.96 (0.34)
12	2.13 (0.34)	2.03 (0.31)	1.95 (0.33)	1.89 (0.35)	1.84 (0.31)	1.79 (0.33)	1.77 (0.33)	1.75 (0.35)
13	1.97 (0.34)	1.63 (0.32)	1.53 (0.30)	1.58 (0.31)	1.69 (0.33)	1.78 (0.30)	1.80 (0.30)	1.75 (0.32)
14	1.83 (0.33)	1.52 (0.34)	1.68 (0.33)	1.80 (0.32)	1.67 (0.34)	1.56 (0.32)	1.68 (0.31)	1.84 (0.32)
15	1.72 (0.34)	1.62 (0.31)	1.80 (0.31)	1.57 (0.31)	1.66 (0.33)	1.84 (0.30)	1.62 (0.30)	1.67 (0.33)
16	1.64 (0.31)	1.78 (0.34)	1.61 (0.30)	1.68 (0.33)	1.77 (0.32)	1.58 (0.31)	1.84 (0.31)	1.61 (0.30)
17	1.58 (0.32)	1.86 (0.34)	1.54 (0.32)	1.82 (0.32)	1.58 (0.31)	1.79 (0.33)	1.64 (0.34)	1.77 (0.31)
18	1.55 (0.35)	1.80 (0.31)	1.77 (0.31)	1.54 (0.30)	1.83 (0.32)	1.64 (0.32)	1.66 (0.34)	1.84 (0.32)
19	1.53 (0.33)	1.67 (0.34)	1.89 (0.34)	1.74 (0.31)	1.54 (0.32)	1.71 (0.33)	1.84 (0.34)	1.67 (0.31)
20	1.52 (0.30)	1.56 (0.32)	1.70 (0.34)	1.85 (0.30)	1.88 (0.31)	1.75 (0.31)	1.61 (0.33)	1.61 (0.31)
21	1.52 (0.30)	1.52 (0.33)	1.53 (0.34)	1.57 (0.30)	1.62 (0.30)	1.70 (0.33)	1.77 (0.31)	1.84 (0.31)
22	3.67 (0.32)	3.50 (0.34)	3.36 (0.31)	3.25 (0.34)	3.16 (0.32)	3.10 (0.34)	3.06 (0.31)	3.04 (0.31)
23	3.40 (0.33)	2.83 (0.34)	2.65 (0.33)	2.73 (0.33)	2.92 (0.30)	3.08 (0.32)	3.10 (0.34)	3.02 (0.30)
24	3.17 (0.32)	2.64 (0.31)	2.91 (0.30)	3.11 (0.31)	2.89 (0.30)	2.70 (0.33)	2.90 (0.34)	3.17 (0.31)
25	2.98 (0.32)	2.81 (0.32)	3.10 (0.31)	2.72 (0.31)	2.87 (0.34)	3.18 (0.32)	2.82 (0.30)	2.89 (0.33)
26	2.84 (0.30)	3.08 (0.32)	2.80 (0.30)	2.90 (0.32)	3.07 (0.30)	2.73 (0.34)	3.18 (0.32)	2.79 (0.33)
27	2.74 (0.30)	3.21 (0.33)	2.67 (0.32)	3.14 (0.34)	2.73 (0.31)	3.10 (0.34)	2.83 (0.31)	3.06 (0.31)
28	2.68 (0.30)	3.11 (0.30)	3.06 (0.33)	2.68 (0.33)	3.16 (0.32)	2.84 (0.32)	2.88 (0.34)	3.17 (0.34)
29	2.65 (0.33)	2.89 (0.34)	3.26 (0.34)	3.01 (0.30)	2.68 (0.30)	2.96 (0.30)	3.17 (0.34)	2.89 (0.33)
30	2.64 (0.31)	2.70 (0.31)	2.94 (0.30)	3.20 (0.32)	3.24 (0.30)	3.02 (0.34)	2.79 (0.33)	2.79 (0.30)
31	2.63 (0.32)	2.64 (0.32)	2.66 (0.33)	2.72 (0.33)	2.82 (0.31)	2.94 (0.32)	3.07 (0.33)	3.18 (0.31)

Rules are defined as given in Table 4.4

Quantities in parentheses represent standard deviation

detect and isolate faults from measured frequencies. We can note that the rules are very complicated and would be difficult to develop manually.

A sample MATLAB code for training the GFS is given in file "Training_code_for_GFS.m", for which the input file is "model2.inp" and the output file is "model2.out".

4.3.2 Testing of Genetic Fuzzy System

The GFS for detecting crack density and location of matrix cracking is tested for a range of constraining angles and numbers of 90° plies. Two cases are discussed in detail, and the results of the other cases are summarized.

1. Constraining angle (θ) for $[\pm\theta_8/90_9]_s$ family of composites.
2. Number of 90° plies (n) for $[\pm30_m/90_n]_s$ family of composites.

The above cases are selected to study the effect of the constraining angle (θ) and the effect of various numbers of 90° plies (n) or the effect of the depth of the matrix crack on the matrix crack density and location prediction.

In the first case, the success rate is calculated for a constraining angle θ of 0, 30, and 60 degrees for the $[\pm\theta_8/90_9]_s$ family of composites. Success rates for these composites for noise levels of 0.05 (below training noise level), 0.1 (training noise level), and 0.15 (above training noise level) are shown in Table 4.4. From this table, it is seen that for all the three layups the success rate is 100% for a noise level of 0.05. As the noise level increases, the quality of the measurement data starts deteriorating and data for one rule can become identified with another; hence, the success rate decreases. For the $[\pm60_8/90_9]_s$ composite, the rules are well separated and therefore the success rate is 100% even for a noise level of 0.15. For the $[0_{16}/90_9]_s$ and $[\pm30_8/90_9]_s$ composites some of the rules become confused for noise levels of 0.1 and 0.15. It is also observed that the success rate is less than 95% only for rules 2 to 11 (slight damage level). The reasons for this can be observed from Fig. 4.3, which shows a greater variation in stiffness, and therefore frequencies, for the $\theta = 60°$ plies than for the $\theta = 0°$ and $\theta = 30°$ plies. The $\theta = 0°$ plies show the least variation in stiffness and therefore the lowest average success rate of 89.4%. The important point here is the 100% success rate for all cases when the noise level is 0.05, which is representative of good data. For a noise level of 0.1, the average success rate lies between 97–100%, which is also very good. For a noise level of 0.15, which is low quality data, the average success rate is 89–100%. For moderate and severe damage levels, the success rate of damage detection is very good for this system.

In the second case, the success rate is calculated for various numbers of 90° plies (n) for the$[\pm30_m/90_n]_s$ family of composites. The success rates for this case at noise levels of 0.05, 0.1, and 0.15 are shown in Table 4.5. It is observed that the success rate is 100% for 17 90° plies. Also, for nine 90° plies, the success rate for all rules is more than 99%, and the average success rate is up to 99.88%. For five 90° plies, the success rate at noise level 0.05 is about 90% for undamaged and slight damage level rules (rule 2 to rule 11), and the success rate is more than 99% for the two other levels of damage. However, for low numbers of the 90° plies, the success rate is not good for slight damage at high noise levels. Observing Fig. 4.3, we can see the reasons behind the behavior of the fuzzy system. As the number of 90° plies decreases, the variation and magnitude of the change in stiffness, and therefore the frequency, decreases. Therefore, the rules show more confusion.

Table 4.4 For the typical $[\pm\theta_8/90_9]_s$ family of composites, success rate at various noise levels and for various constraining angles

No.	Rule	θ								
		0°			30°			60°		
		$S_{R0.05}$	$S_{R0.1}$	$S_{R0.15}$	$S_{R0.05}$	$S_{R0.1}$	$S_{R0.15}$	$S_{R0.05}$	$S_{R0.1}$	$S_{R0.15}$
1	Undamaged	100.00	100.00	100.00	100.00	100.00	100.00	100.00	100.00	100.00
2	Slight L_{10}	100.00	100.00	97.50	100.00	100.00	100.00	100.00	100.00	100.00
3	Slight L_{20}	100.00	92.90	71.70	100.00	99.80	90.40	100.00	100.00	100.00
4	Slight L_{30}	100.00	91.10	63.90	100.00	99.70	87.80	100.00	100.00	100.00
5	Slight L_{40}	100.00	89.90	61.20	100.00	99.50	82.20	100.00	100.00	100.00
6	Slight L_{50}	100.00	84.00	53.20	100.00	98.80	82.30	100.00	100.00	100.00
7	Slight L_{60}	100.00	93.60	65.80	100.00	99.80	89.50	100.00	100.00	100.00
8	Slight L_{70}	100.00	90.60	66.80	100.00	99.10	87.10	100.00	100.00	100.00
9	Slight L_{80}	100.00	88.90	65.20	100.00	99.70	88.50	100.00	100.00	100.00
10	Slight L_{90}	100.00	93.70	74.10	100.00	100.00	90.30	100.00	100.00	100.00
11	Slight L_{100}	100.00	91.80	68.00	100.00	99.90	89.50	100.00	100.00	100.00
12	Moderate L_{10}	100.00	100.00	100.00	100.00	100.00	100.00	100.00	100.00	100.00
13	Moderate L_{20}	100.00	100.00	98.60	100.00	100.00	100.00	100.00	100.00	100.00
14	Moderate L_{30}	100.00	100.00	99.00	100.00	100.00	99.80	100.00	100.00	100.00
15	Moderate L_{40}	100.00	100.00	97.10	100.00	100.00	99.40	100.00	100.00	100.00
16	Moderate L_{50}	100.00	100.00	95.80	100.00	100.00	98.50	100.00	100.00	100.00
17	Moderate L_{60}	100.00	100.00	99.20	100.00	100.00	99.90	100.00	100.00	100.00
18	Moderate L_{70}	100.00	100.00	98.20	100.00	100.00	99.30	100.00	100.00	100.00
19	Moderate L_{80}	100.00	100.00	98.30	100.00	100.00	100.00	100.00	100.00	100.00
20	Moderate L_{90}	100.00	100.00	99.30	100.00	100.00	100.00	100.00	100.00	100.00
21	Moderate L_{100}	100.00	100.00	98.50	100.00	100.00	100.00	100.00	100.00	100.00
22	Severe L_{10}	100.00	100.00	100.00	100.00	100.00	100.00	100.00	100.00	100.00
23	Severe L_{20}	100.00	100.00	100.00	100.00	100.00	100.00	100.00	100.00	100.00
24	Severe L_{30}	100.00	100.00	100.00	100.00	100.00	100.00	100.00	100.00	100.00
25	Severe L_{40}	100.00	100.00	100.00	100.00	100.00	100.00	100.00	100.00	100.00
26	Severe L_{50}	100.00	100.00	100.00	100.00	100.00	100.00	100.00	100.00	100.00
27	Severe L_{60}	100.00	100.00	100.00	100.00	100.00	100.00	100.00	100.00	100.00
28	Severe L_{70}	100.00	100.00	100.00	100.00	100.00	100.00	100.00	100.00	100.00
29	Severe L_{80}	100.00	100.00	100.00	100.00	100.00	100.00	100.00	100.00	100.00
30	Severe L_{90}	100.00	100.00	100.00	100.00	100.00	100.00	100.00	100.00	100.00
31	Severe L_{100}	100.00	100.00	100.00	100.00	100.00	100.00	100.00	100.00	100.00
	Avg S_R	100.00	97.31	89.40	100.00	99.88	96.27	100.00	100.00	100.00

An important point to note from these results is the 100% correct classification of the "Undamaged" condition. It is very important for a diagnostic system not to give false positives, which is when a damage is indicated by the system when none

Table 4.5 For the typical $[\pm30_n/90_9]_s$ family of composites, success rate at various noise levels and various numbers of 90° plies

No.	Rule	n								
		5			9			17		
		$S_{R0.05}$	$S_{R0.1}$	$S_{R0.15}$	$S_{R0.05}$	$S_{R0.1}$	$S_{R0.15}$	$S_{R0.05}$	$S_{R0.1}$	$S_{R0.15}$
1	Undamaged	100.00	100.00	100.00	100.00	100.00	100.00	100.00	100.00	100.00
2	Slight L_{10}	100.00	87.00	67.60	100.00	100.00	100.00	100.00	100.00	100.00
3	Slight L_{20}	94.60	51.30	36.40	100.00	99.80	90.40	100.00	100.00	100.00
4	Slight L_{30}	89.80	43.80	24.20	100.00	99.70	87.80	100.00	100.00	100.00
5	Slight L_{40}	87.60	39.40	27.70	100.00	99.50	82.20	100.00	100.00	100.00
6	Slight L_{50}	86.30	38.30	26.00	100.00	98.80	82.30	100.00	100.00	100.00
7	Slight L_{60}	92.00	52.00	33.70	100.00	99.80	89.50	100.00	100.00	100.00
8	Slight L_{70}	89.70	48.70	30.20	100.00	99.10	87.10	100.00	100.00	100.00
9	Slight L_{80}	89.60	46.40	34.90	100.00	99.70	88.50	100.00	100.00	100.00
10	Slight L_{90}	92.80	55.10	37.40	100.00	100.00	90.30	100.00	100.00	100.00
11	Slight L_{100}	91.00	43.20	29.00	100.00	99.90	89.50	100.00	100.00	100.00
12	Moderate L_{10}	100.00	100.00	95.90	100.00	100.00	87.10	100.00	100.00	100.00
13	Moderate L_{20}	99.80	84.30	59.90	100.00	100.00	88.50	100.00	100.00	100.00
14	Moderate L_{30}	100.00	84.40	61.50	100.00	100.00	90.30	100.00	100.00	100.00
15	Moderate L_{40}	99.70	68.80	35.80	100.00	100.00	89.50	100.00	100.00	100.00
16	Moderate L_{50}	99.30	66.50	37.30	100.00	100.00	90.30	100.00	100.00	100.00
17	Moderate L_{60}	99.90	86.30	57.40	100.00	100.00	100.00	100.00	100.00	100.00
18	Moderate L_{70}	99.50	81.50	58.50	100.00	100.00	100.00	100.00	100.00	100.00
19	Moderate L_{80}	99.80	87.50	66.40	100.00	100.00	99.80	100.00	100.00	100.00
20	Moderate L_{90}	100.00	85.40	60.60	100.00	100.00	99.40	100.00	100.00	100.00
21	Moderate L_{100}	100.00	85.40	56.70	100.00	100.00	98.50	100.00	100.00	100.00
22	Severe L_{10}	100.00	100.00	100.00	100.00	100.00	99.90	100.00	100.00	100.00
23	Severe L_{20}	100.00	100.00	96.90	100.00	100.00	99.30	100.00	100.00	100.00
24	Severe L_{30}	100.00	99.90	90.30	100.00	100.00	100.00	100.00	100.00	100.00
25	Severe L_{40}	100.00	99.50	85.00	100.00	100.00	100.00	100.00	100.00	100.00
26	Severe L_{50}	100.00	97.10	74.10	100.00	100.00	100.00	100.00	100.00	100.00
27	Severe L_{60}	100.00	99.70	95.70	100.00	100.00	100.00	100.00	100.00	100.00
28	Severe L_{70}	100.00	98.40	89.30	100.00	100.00	100.00	100.00	100.00	100.00
29	Severe L_{80}	100.00	99.70	95.30	100.00	100.00	100.00	100.00	100.00	100.00
30	Severe L_{90}	100.00	100.00	96.60	100.00	100.00	100.00	100.00	100.00	100.00
31	Severe L_{100}	100.00	100.00	96.70	100.00	100.00	100.00	100.00	100.00	100.00
	Avg S_R	97.31	78.40	63.13	100.00	99.88	94.52	100.00	100.00	100.00

is present. The case of misclassification is of relatively lesser importance than false positives which can severely damage the credibility of the diagnostic system and lead to high costs. Also, the GFS serves the function of both fault detection and

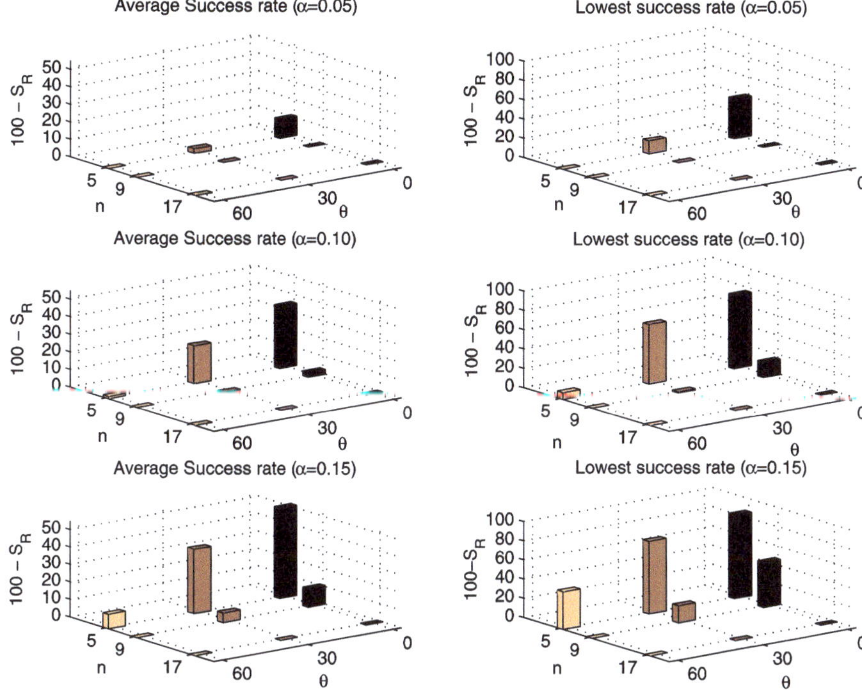

Fig. 4.5 Average success rate and lowest success rate of GFS for various combinations of $[\pm\theta_m/90_n]_s$ family of composites

isolation. If rule 1 is active, there is no damage; if the other rules are active, there is damage present in the structure.

Finally, we summarize the results of testing the composite beam for damage detection at other combinations of θ and n. Figure 4.5 shows the behavior of the average success rate and the lowest success rate for the three noise levels and various combinations of the $[\pm\theta_m/90_n]_s$ family of composites. From this figure, the performance of the GFS and the reliability zone of this system can be determined. The performance is best for $\theta = 60°$ and $n = 17$ composites. As n decreases to 5 and θ decreases to 0°, the performance of correctly locating the damage decreases. However, as we shall see in the next section, the damage detection accuracy (is a damage present or not?) and identification capability (how much damage is present?) remain high.

A sample MATLAB code for testing the GFS is given in file "Testing_code_for_GFS.m", for which input files are "rul309.dat" and "tune.dat", whereas the output file is "f09_15.dat".

4.3.3 Analysis of Misclassification

In some cases of highly noisy data, the rules which are not correctly classified lead to a success rate of less than 100%. If there is a misclassification, it is important

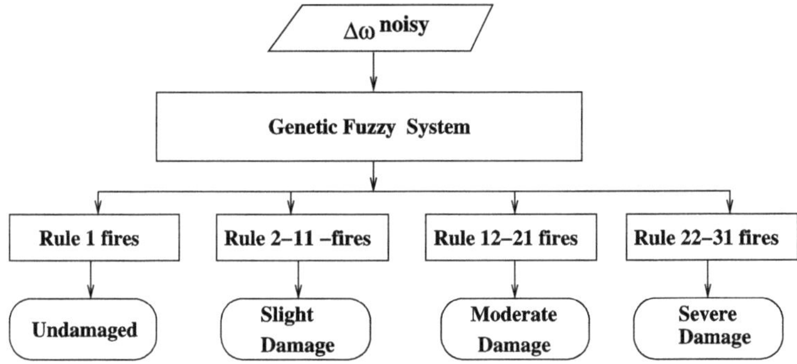

Fig. 4.6 Damage detection and identification using GFS

to know if the choice is slightly wrong or completely wrong. Thus, if the system gives a neighboring location of damage, it is much less harmful than if it gives a distant location. A diagnostic system should gradually deteriorate as data quality deteriorates; it should not show a sudden fall in performance.

For example, rules 3 to 11 for the $[0_{16}/90_9]_s$ composite at a noise level of 0.15 are considered whose success rates are less than 75%, as shown in Table 4.4. In these cases, more than 25% of the data is distributed in other rules. It is observed that wrongly classified rules fall in the same level of damage but at the wrong location. From Table 4.4, rules 2 to 11 all indicate "Slight" damage level. Therefore, in terms of damage level detection, the fuzzy system is able to classify all cases of "Slight" damage level with an accuracy of 100%. The same results are also found in the cases for "Moderate" and "Severe" damage level. Therefore, the fuzzy system is robust in terms of indicating the size of the damage even at high levels of noise in the data, although there can be some error in locating the damage. Figure 4.6 shows a schematic representation of how the fuzzy system can be used for the detection and identification of damage. Note that the accuracy of these functions is very good ($\approx$100) for the GFS, even for low quality data, while location accuracy can deteriorate in certain situations [11].

Some MATLAB code illustrating the process of SHM development using the GFS is given at the end of the book. A detailed description of the code is given in Appendix.

4.4 Summary

In this chapter, a genetic fuzzy system is developed and demonstrated for structural damage detection of a thin-walled composite tube. Such thin-walled composite structures are widely used in applications such as electrical transmission poles. The damage modeled is matrix cracking, and the damage indicators are the changes in natural frequencies relative to the undamaged structure. Localized matrix cracking is typically initiated in composites by low velocity impact. The genetic fuzzy system is developed by maximizing the success rate of damage detection and tuning the fuzzy sets and rules using genetic algorithms. The genetic fuzzy system predicts the matrix cracking accurately for reasonable noise levels in the measurement delta. It should be noted that the damage detection approach in this chapter is developed using a finite element model and is also tested using simulated data obtained by adding noise to finite element-based predictions. The algorithm developed in this chapter provides a new approach toward achieving a user-friendly damage detection system which provides linguistic and qualitative information from complicated and noisy numerical data. In the next chapter, the genetic fuzzy system will be used for online health monitoring of a helicopter rotor using simulated measurements from an aeroelastic simulation. Both the structural and the damage model complexity will increase substantially.

References

1. Saboori, B., Khalili, S.M.R.: Static analysis of tapered FRP transmission poles using finite element method. Finite Elem. Anal. Des. **47**(3), 247–255 (2011)
2. Raftoyiannis, I.G., Polyzois, D.J.: The effect of semi-rigid connections on the dynamic behavior of tapered composite GRP poles. Compos. Struct. **81**(1), 70–79 (2007)
3. Polyzois, D.J., Raftoyiannis, I.G., Ungkurapinan, N.: Static and dynamic characteristics of multi-cell jointed GFRP wind turbine towers. Compos. Struct. **90**(1), 34–42 (2009)
4. Kim, J.S., Yoon, H.J., Shin, K.B.: A study of crushing behaviors of composite circular tubes with different reinforcing fibers. Int. J. Impact Eng. **38**(4), 198–207 (2011)
5. Jones, R.M.: Mechanics of Composite Materials. Taylor & Francis, New York (1999)
6. Ferrero, J.F., Barrau, J.J., Segura, J.M., Sudre, M., Castanie, B.: Analytical theory for an approach calculation of non-balanced composite box beams. Thin-Walled Struct. **39**, 709–29 (2001)
7. Polyzois, D., Raftoyiannis, G., Ibrahim, S.: Finite elements for the dynamic analysis of tapered composite poles. Compos. Struct. **43**, 25–34 (1998)
8. Estivalezes, E., Barrau, J.J.: Analytical theory for an approach calculation of composite box beams subjected to tension and bending. Compos., Part B **29**, 371–376 (1998)
9. Nuismer, R.J., Tan, S.C.: Constitutive relations of a cracked composite lamina. J. Compos. Mater. **22**, 306–21 (1988)
10. Tiberkak, R., Bachene, M., Rechak, S., Necib, B.: Damage prediction in composite plates subject to low velocity impact. Compos. Struct. **83**(1), 73–82 (2008)
11. Pawar, P.M., Ganguli, R.: Matrix crack detection in thin-walled composite beam using genetic fuzzy system. J. Intell. Mater. Syst. Struct. **16**(5), 395–409 (2005)

Chapter 5
Structural Health Monitoring of Composite Helicopter Rotor

The last two chapters have shown the application of the genetic fuzzy system to beams and tubes made of different materials. While the beam is an important structural member, the composite tube represents a real structure such as a transmission pole. In this chapter, we develop the genetic fuzzy system for structural health monitoring (SHM) of a composite helicopter rotor blade. Such a system is deployed for detecting damage in the composite rotor. Mathematical modeling of the damaged helicopter rotor is required for developing the rule base for the genetic fuzzy system. The mathematical models of composite damage also become quite elaborate, as we need to simulate the damaged system in real life. Furthermore, modeling of the helicopter rotor blade requires an aeroelastic analysis, as the highly flexible blades cause the structure and aerodynamics to be coupled. For example, a damage in the rotor blade which causes a change in the blade twist will also affect the aerodynamic loads due to a change in the blade section angle of attack. Therefore, damage affects the system properties of the helicopter rotor in a complicated and often nonlinear manner. The case study presented in this chapter is therefore for a complex system whose modeling (both the undamaged and damaged system) represents a high level of complexity.

The helicopter rotor blades operate in a highly dynamic and unsteady aerodynamic environment leading to severe vibratory loads on the rotor system. Repeated exposure to this severe loading condition can induce damage in the composite rotor blade. These blades, generally made of fiber reinforced laminated composites, exhibit various competing damage modes such as matrix cracking, delamination, fiber matrix debonding, fiber breakage, and fiber pull-out.

Generally, as mentioned in the previous chapter, the first damage mode in a composite material is dominated by matrix cracking. The number of cracks increases monotonically with load or with the number of load cycles until a saturation density is reached. Due to the interlaminar cracks, separation of the plies locally takes place. The separation of plies is known as interior delamination, as opposed to the exterior delamination associated with free edges in laminates. The final failure mode is fiber breakage, which is associated with the ultimate failure of the ply. Under cyclic loading, the accumulation of damage in the rotor blade follows a systematic process. First, the matrix cracks accumulate and then reach saturation after some time.

P.M. Pawar, R. Ganguli, *Structural Health Monitoring Using Genetic Fuzzy Systems*,
DOI 10.1007/978-0-85729-907-9_5, © Springer-Verlag London Limited 2011

Second, the matrix crack saturation is followed by the growth of delamination in the material. Third, fiber breaks start taking place in the composite structure, leading to complete failure of the laminate.

5.1 Mathematical Model

Just as in the previous two chapters for beams and composite tubes, development of an SHM system requires a mathematical model for the undamaged baseline structure and the different cases of the damaged structure. Such a model allows numerical experiments which simulate the damaged helicopter rotor in flight. Obviously, it is not feasible to seed various damages in a composite helicopter rotor and conduct flight tests due to safety concerns and airworthiness requirements. Modeling is therefore indispensable for the health monitoring of aerospace structures.

The analysis of the damaged composite rotor blade is explained in three parts. The first part summarizes the mathematical model of a baseline undamaged helicopter rotor system. The second part discusses the composite rotor blade cross-sectional analysis. The composite cross-sectional properties are included in the mathematical model of the helicopter rotor system. The third part discusses the modeling of the key damage modes in composite materials. The damage modeling is done in a manner such that it can be easily integrated into the composite cross-sectional model. An outline of the formulation of the forward flight simulation of the damaged composite rotor blade is now given. This mathematical model is coded into a computer program which is more than 50,000 lines long. The mathematical model is based on the finite element method, which is generally the norm in the analysis of complicated and realistic structures.

5.1.1 Mathematical Model of Helicopter Rotor

The helicopter is represented by a nonlinear model of rotating elastic rotor blades dynamically coupled to a six-degree-of-freedom rigid fuselage. Each blade undergoes flap bending, lag bending, elastic twist, and axial displacement. Here, flap bending is also called out-of-plane bending and lag bending is called in-plane bending. The elastic twist motion is also called torsion. Governing partial differential equations for the generalized beam are derived using a generalized Hamilton's principle applicable to non-conservative systems [1]:

$$\int_{\psi_1}^{\psi_2} (\delta U - \delta T - \delta W)\, d\psi = 0. \tag{5.1}$$

δU, δT, and δW are the virtual strain energy, kinetic energy, and virtual work, respectively. The azimuth angle around the rotor disk is denoted as ψ, and ψ_1 and ψ_2 represent the initial and final values of these angles. The δU and δT include energy contributions from components that are attached to the blade. These equations are

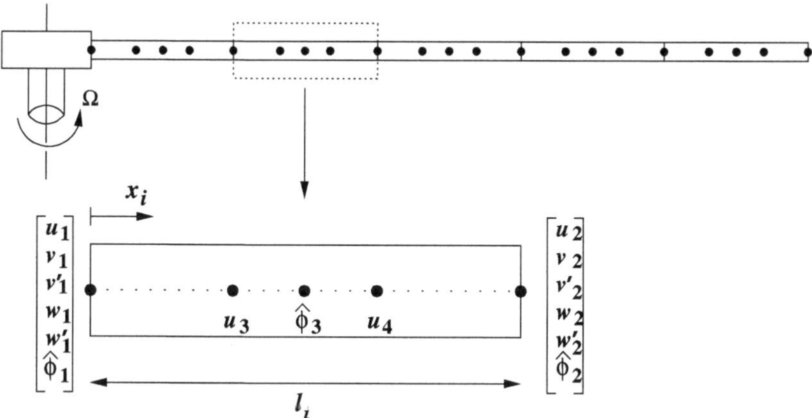

Fig. 5.1 Beam finite element used for rotor blade

based on the work of Hodges and Dowell [1] and include second-order geometric nonlinear terms accounting for moderate blade deflections in the flap bending, lag bending, axial, and torsion equations. External aerodynamic forces on the blade contribute to the virtual work variational, δW. The aerodynamic forces and moments are calculated using unsteady aerodynamics [2]. The finite element method is used to discretize the governing equations of motion, and allows for accurate representation of complex hub kinematics and nonuniform blade properties [3]. After the finite element discretization into N elements, Hamilton's principle is written as

$$\int_{\psi_i}^{\psi_f} \sum_{i=1}^{N} (\delta U_i - \delta T_i - \delta W_i)\, d\psi = 0. \tag{5.2}$$

Each of the N beam elements has fifteen degrees of freedom, as shown in Fig. 5.1.

These degrees of freedom are distributed over five element nodes (two boundary nodes and three interior nodes). There are six degrees of freedom at each element boundary node. These six degrees of freedom correspond to u, v, v', w, w', and $\widehat{\phi}$. There are two internal nodes for axial deflection u and one internal node for elastic twist $\widehat{\phi}$. There is continuity of displacements and slope for flap (w) and lag (v) bending deflections between elements. Also, there is inter-element continuity of displacement for elastic twist and axial deflections. These elements ensure physically consistent linear variations of bending moments and torsional moment, and quadratic variation of axial force within each element. Using the interpolating polynomials, the distribution of deflections over a beam element is expressed in terms of the elemental nodal displacements $\mathbf{q}_i$. The shape functions used are Hermite polynomials for lag and flap bending, and Lagrange polynomials for axial and torsion deflection [4]. For the ith beam element, the elemental nodal displacement vector is defined as

$$\mathbf{q}_i^T = [u_1, u_2, u_3, u_4, v_1, v_1', v_2, v_2', w_1, w_1', w_2, w_2', \hat{\phi}_1, \hat{\phi}_2, \hat{\phi}_3]. \tag{5.3}$$

Assembling the blade finite element equations and applying boundary conditions results in (5.2) becoming [3, 5]

$$\mathbf{M}\ddot{\mathbf{q}}(\psi) + \mathbf{C}\dot{\mathbf{q}}(\psi) + \mathbf{K}\mathbf{q}(\psi) = \mathbf{F}(\mathbf{q}, \dot{\mathbf{q}}, \psi). \tag{5.4}$$

Here $\mathbf{M}$, $\mathbf{C}$, and $\mathbf{K}$ represent the mass, damping, and stiffness matrices for the rotor blade and $\mathbf{F}$ is the force vector. The nodal displacements $\mathbf{q}$ are functions of time, and all nonlinear terms have been moved into the force vector on the right-hand side. The spatial functionality has been removed by using finite element discretization, and partial differential equations have been converted into ordinary differential equations (ODEs).

The finite element equations representing each rotor blade are transformed to normal mode space for efficient solution of blade response using the modal expansion. Typically, structural dynamic systems with a very large number of degrees of freedom can be often represented by using the first few normal modes, resulting in greatly enhanced computational efficiency. Typically, six to ten modes are found to be sufficient for capturing the helicopter dynamics. The displacements are expressed in terms of normal modes as

$$\mathbf{q} = \Phi \mathbf{p}. \tag{5.5}$$

Here, $\mathbf{p}$ is the vector of modal coordinates and Φ contains the eigenvectors corresponding to free vibration of the rotor blade. Substituting (5.5) into (5.4) lead to normal mode equations having the form

$$\bar{\mathbf{M}}\ddot{\mathbf{p}}(\psi) + \bar{\mathbf{C}}\dot{\mathbf{p}}(\psi) + \bar{\mathbf{K}}\mathbf{p}(\psi) = \bar{\mathbf{F}}(\mathbf{p}, \dot{\mathbf{p}}, \psi). \tag{5.6}$$

These equations are nonlinear ODEs, but their dimensions are much reduced compared to the full finite element equations (5.4). The normal mode mass, stiffness, and damping matrices and force vectors are given by

$$\bar{\mathbf{M}} = \Phi^T \mathbf{M} \Phi, \qquad \bar{\mathbf{C}} = \Phi^T \mathbf{C} \Phi, \qquad \bar{\mathbf{K}} = \Phi^T \mathbf{K} \Phi, \qquad \bar{\mathbf{F}} = \Phi^T \mathbf{F}. \tag{5.7}$$

The mode shapes or eigenvectors in (5.5) and (5.7) are obtained by solving the general eigenvalue problem [3]

$$\mathbf{K}_s \Phi = \omega^2 \mathbf{M}_s \Phi. \tag{5.8}$$

Here, $\mathbf{K}_s$ and $\mathbf{M}_s$ represent the global stiffness and mass matrices for free vibration of the rotor blade, and ω represents the rotating frequencies of the blade. Note that rotating blades have a frequency which includes the effect of centrifugal stiffening, and so their frequencies are greater than those of non-rotating blades. In fact, non-rotating blades are well modeled as beams, as was done in Chap. 3 for the BO105 rotor blade with only the flapwise mode.

The blade normal mode equations (5.6) can be written in the following variational form [7]:

$$\int_0^{2\pi} \delta \mathbf{p}^T (\bar{\mathbf{M}}\ddot{\mathbf{p}} + \bar{\mathbf{C}}\dot{\mathbf{p}} + \bar{\mathbf{K}}\mathbf{p} - \bar{\mathbf{F}})\, d\psi = 0. \tag{5.9}$$

Integrating (5.9) by parts, we obtain [7]

$$\int_0^{2\pi} \begin{Bmatrix} \delta \mathbf{p} \\ \delta \dot{\mathbf{p}} \end{Bmatrix} \begin{Bmatrix} \bar{\mathbf{F}} - \bar{\mathbf{C}}\dot{\mathbf{p}} - \bar{\mathbf{K}}\mathbf{p} \\ \bar{\mathbf{M}}\dot{\mathbf{p}} \end{Bmatrix} d\psi = \begin{Bmatrix} \delta \mathbf{p} \\ \delta \dot{\mathbf{p}} \end{Bmatrix} \begin{Bmatrix} \mathbf{M}\dot{\mathbf{p}} \\ \mathbf{0} \end{Bmatrix} \Bigg|_0^{2\pi}. \tag{5.10}$$

Since the helicopter rotor is a periodic system with a time period of one revolution, we have $\dot{\mathbf{p}}(0) = \dot{\mathbf{p}}(2\pi)$. Imposing periodic boundary conditions on (5.10) results in the right-hand side becoming zero and yields the following system of first-order ODEs [7]:

$$\int_0^{2\pi} \delta \mathbf{y}^T \mathbf{Q}\, d\psi = 0, \tag{5.11}$$

where

$$\mathbf{y} = \begin{Bmatrix} \mathbf{p} \\ \dot{\mathbf{p}} \end{Bmatrix}, \qquad \mathbf{Q} = \begin{Bmatrix} \bar{\mathbf{F}} - \bar{\mathbf{C}}\dot{\mathbf{p}} - \bar{\mathbf{K}}\mathbf{p} \\ \bar{\mathbf{M}}\dot{\mathbf{p}} \end{Bmatrix}. \tag{5.12}$$

The nonlinear, periodic ODEs are then solved for blade steady response using the finite element in time method and a Newton–Raphson procedure [7]. Discretizing (5.12) over N_t time elements around the rotor disk (where $\psi_1 = 0$, $\psi_{N_t+1} = 2\pi$) and taking a first-order Taylor series expansion about the steady-state value $\mathbf{y}_o = [\mathbf{p}_o^T \dot{\mathbf{p}}_o^T]$ yields algebraic equations [7]

$$\sum_{i=1}^{N_i} \int_{\psi_i}^{\psi_{i+1}} \delta \mathbf{y}_i^T Q_i(\mathbf{y}_o + \Delta \mathbf{y})\, d\psi$$

$$= \sum_{i=1}^{N_i} \int_{\psi_i}^{\psi_{i+1}} \delta \mathbf{y}_i^T \left[Q_i(\mathbf{y}_o) + K_{ti}(\mathbf{y}_o)\Delta \mathbf{y} \right] d\psi = 0, \tag{5.13}$$

where the tangential stiffness matrix is given by

$$\mathbf{K}_{ti} = \begin{bmatrix} \frac{\partial \bar{F}}{\partial P} - \bar{\mathbf{K}} & \frac{\partial \bar{F}}{\partial \dot{P}} - \bar{\mathbf{C}} \\ 0 & \bar{\mathbf{M}} \end{bmatrix}_i. \tag{5.14}$$

For the ith time element, the modal displacement vector can be written as

$$\mathbf{p}_i(\psi) = \mathbf{H}(s)\xi_i, \tag{5.15}$$

where $\mathbf{H}$(s) are time shape functions which are fifth-order Lagrange polynomials [7] used for approximating the normal mode coordinate $\mathbf{p}$. For a fifth-order polynomial, six nodes are needed to describe the variation of $\mathbf{p}$ within the element. Continuity of generalized displacements is assumed between the time elements. Substituting (5.15) and its derivative into (5.14) yields the time-discretized blade response [7]

$$\mathbf{Q}^G + \mathbf{K}_i{}^G \Delta \xi^G = 0, \tag{5.16}$$

where

$$\mathbf{Q}^G = \sum_{i=1}^{N_i} \int_{\psi_i}^{\psi_{i+1}} \mathbf{H}^T \mathbf{Q}_i \, d\psi, \tag{5.17}$$

$$\mathbf{K}_t^G = \sum_{i=1}^{N_t} \int_{\psi}^{\psi_{i+1}} \mathbf{H}^T \begin{bmatrix} \frac{\partial \bar{F}}{\partial P} - \bar{\mathbf{K}} & \frac{\partial \bar{F}}{\partial P} - \bar{\mathbf{C}} \\ & \bar{\mathbf{M}} \end{bmatrix}_i d\psi,$$
$$\Delta\xi^G = \sum_{i=1}^{N_t} \Delta\xi_i. \tag{5.18}$$

Solving the above equations iteratively yields the blade steady response.

Steady and vibratory components of the rotating frame blade loads (i.e., shear forces and bending/torsion moments) are calculated using the force summation method. In this approach, blade inertia and aerodynamic forces are integrated directly over the length of the blade. The blade root loads are given as [5]

$$\begin{Bmatrix} F_{xR} \\ F_{yR} \\ F_{zR} \end{Bmatrix} = \int_0^1 \begin{Bmatrix} L_u \\ L_v \\ L_w \end{Bmatrix} dx, \tag{5.19}$$

$$\begin{Bmatrix} M_{xR} \\ M_{yR} \\ M_{zR} \end{Bmatrix} = \int_0^1 \begin{Bmatrix} -L_v w + L_w v + M_u \\ L_u w - L_w v + M_v \\ -L_u v + L_v(x+u) + M_w \end{Bmatrix} dx. \tag{5.20}$$

Here L_u, L_v, and L_w represent the blade section forces and M_u, M_v, and M_w represent the blade section moments. Fixed frame hub loads are calculated by summing the individual contributions of individual blades [5]:

$$F_x^H(\psi) = \sum_{m=1}^{N_b} \left(F_x^m \cos\psi_m - F_y^m \sin\psi_m - \beta_p F_z^m \cos\psi_m\right),$$
$$F_y^H(\psi) = \sum_{m=1}^{N_b} \left(F_x^m \sin\psi_m + F_y^m \cos\psi_m - \beta_p F_z^m \sin\psi_m\right),$$
$$F_z^H(\psi) = \sum_{m=1}^{N_b} \left(F_z^m + \beta_p F_x^m\right),$$
$$M_x^H(\psi) = \sum_{m=1}^{N_b} \left(M_x^m \cos\psi_m - M_y^m \sin\psi_m - \beta_p M_z^m \cos\psi_m\right),$$
$$M_y^H(\psi) = \sum_{m=1}^{N_b} \left(M_x^m \sin\psi_m + M_y^m \cos\psi_m - \beta_p M_z^m \sin\psi_m\right),$$
$$M_z^H(\psi) = \sum_{m=1}^{N_b} \left(M_z^m + \beta_p M_x^m\right). \tag{5.21}$$

Here, N_b is the number of rotor blades, β_p is the precone angle which is built in to reduce the flap displacement, and ψ_m is the azimuth angle for the mth blade. Once the hub loads are obtained, the helicopter needs to be trimmed. This is defined as the condition where the steady forces and moments acting on the helicopter sum to zero and simulates the condition for steady level flight [6]. The trim solution for the helicopter involves finding the pilot control angles $\boldsymbol{\theta}$ at which the six steady forces and moments acting on the helicopter are zeros:

$$\mathbf{F}(\boldsymbol{\theta}) = \mathbf{0}. \tag{5.22}$$

Here $\boldsymbol{\theta} = [\theta_0\ \theta_{1c}\ \theta_{1s}\ \alpha_s\ \phi_s\theta_T]$. The angles θ_0, θ_{1c}, θ_{1s}, and θ_T are the four helicopter controls used by the pilot called the collective pitch, lateral cyclic, longitudinal cyclic, and tail rotor collective, respectively. Also, α_s and ϕ_s are the shaft tilt and bank angle of the helicopter, respectively.

The trim equations are solved iteratively using a Newton–Raphson procedure [5]. A coupled trim procedure is carried out to solve the blade response, pilot input trim controls, and vehicle orientation, simultaneously. This procedure is called coupled trim since the blade response equations (5.16) and trim equations (5.22) are simultaneously solved, thereby accounting for the influence of elastic blade deflections on the rotor steady forces [5, 7]:

$$\Delta\theta = -\left.\frac{\partial \mathbf{F}}{\partial \theta}\right|_{\theta_0} (\theta - \theta_0), \tag{5.23}$$

$$\xi^G_{i+1} = \xi^G_{i+1} + \Delta\xi^G_{i+1}. \tag{5.24}$$

The coupled trim is solved iteratively until convergence. The coupled trim procedure is essential for elastically coupled blades since elastic deflections play an important role in the steady net forces and moments generated by the rotor.

Figure 5.2 shows the flowchart of the aeroelastic analysis. Starting with the input to the analysis, structural modeling is carried out as the first step.

The governing differential equations thus obtained are space and time dependent. These equations are nonlinear and periodic in nature. These equations are discretized by using the finite element method in space, and rotating blade natural frequencies and mode shapes are found. The frequency analysis is conducted using linear equations in a vacuum condition assuming a spinning blade ($\theta_0 = \theta_{1c} = \theta_{1s} = 0$). Normal mode transformation is done on the full finite element equations to reduce the number of degrees of freedom and thus reduce the computational time. Aerodynamic modeling is the next major step and is done along with the derivation of the finite element equations. Here the type of aerodynamic model is decided and the governing equations are discretized in time. Thus, both space and time dependencies are removed, resulting in nonlinear, algebraic equations which are solved using the Newton–Raphson method. Blade loads and the helicopter steady loads are calculated, which are in turn required for performing trim calculations. Blade steady state response, vehicle orientation, and trim control angles are solved for iteratively using a coupled trim procedure. The rotor nonlinear equations are linearized about the trim solution to perform aeromechanical stability.

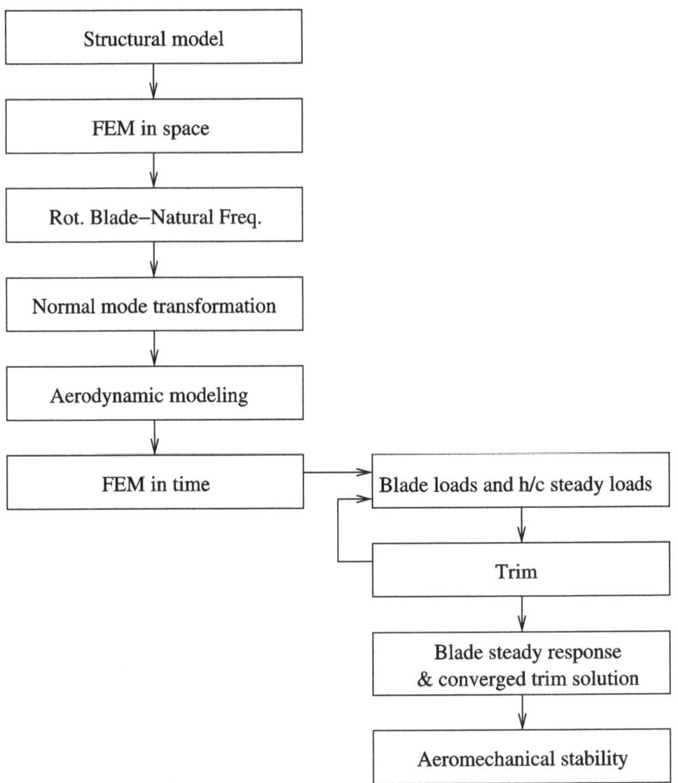

Fig. 5.2 Organization of aeroelastic analysis

5.1.2 Composite Rotor Blade

The previous section discussed the aeroelastic analysis of a rotor blade. The analysis is general in nature, as it uses the energy approach and different materials can be considered in the analysis by adjusting the strain energy component in the derivation. The composite helicopter blade is modeled as a one-dimensional, thin-walled beam undergoing extension, torsion, flap (out-of-plane), and lag (in-plane) bending using the Chandra and Chopra [8] model. The Chandra and Chopra [8] thin-walled composite beam model includes terms due to constrained warping torsion and terms due to transverse shear. The effects of transverse shear are included by static condensation. The restrained warping effect is negligible for a closed section and is therefore ignored for this work. The stiffness matrix of order (9×9) is thus reduced to a stiffness matrix of order (4×4):

$$\begin{bmatrix} N \\ M_x \\ -M_y \\ T_s \end{bmatrix} = \begin{bmatrix} K'_{11} & K'_{12} & K'_{13} & K'_{15} \\ & K'_{22} & K'_{23} & K'_{25} \\ sym. & & K'_{33} & K'_{35} \\ & & & K'_{55} \end{bmatrix} = \begin{bmatrix} W' \\ \phi'_y \\ \phi'_x \\ \phi'_z \end{bmatrix}. \tag{5.25}$$

The coefficients K'_{ij} of the stiffness matrix are obtained by static condensation of the (9×9) stiffness matrix **K**. The terms in the **K** matrix depend on the beam cross section and geometry and are expressed in terms of **A**, **B**, and **D** matrices. Here $EA = K'_{11}$, $EI_y = K'_{22}$, $EI_z = K'_{33}$, $GJ = K'_{55}$ are the axial, flap, lag, and torsion stiffness, respectively. The effects of composite material are included in the forward flight simulation through the strain energy expression using the cross-sectional stiffness matrix derived in this section. The effects of the key damages in composite material are included in the forward flight simulation through the cross-sectional stiffness matrices using the progressive damage accumulation model given in the next section.

5.1.3 Progressive Damage Accumulation

Matrix cracking, delamination/debonding, and fiber breakage are the key damage modes in composite materials. These damage modes are modeled at the lamina and laminate level, as summarized next.

5.1.3.1 Matrix Cracking

The matrix cracking in composite structures is inserted through the extension (**A**), extension-bending (**B**), and bending (**D**) stiffness matrices. The stiffness matrices for the presence of matrix cracks $\mathbf{A}^{(c)}$, $\mathbf{B}^{(c)}$, and $\mathbf{D}^{(c)}$ are obtained by subtracting the damage matrices $\Delta\mathbf{A}$, $\Delta\mathbf{B}$, and $\Delta\mathbf{D}$ from the stiffness matrices **A**, **B**, and **D** of the virgin laminate:

$$\mathbf{A}^{(c)} = \mathbf{A} - \Delta\mathbf{A}, \tag{5.26}$$

$$\mathbf{B}^{(c)} = \mathbf{B} - \Delta\mathbf{B}, \tag{5.27}$$

$$\mathbf{D}^{(c)} = \mathbf{D} - \Delta\mathbf{D}. \tag{5.28}$$

These stiffness matrices reduce with increasing crack density. The dimensionless crack density ρ^k for the ply k is defined by

$$\rho^k = \frac{t^k}{s^k}. \tag{5.29}$$

Here s^k is the average crack spacing and t^k is the thickness of ply k. The **A**, **B**, and **D** matrices are defined in terms of the lamina stiffness for the virgin laminate and are given by

$$\mathbf{A} = \sum_{k=1}^{N} t^k \mathbf{Q}^k, \tag{5.30}$$

$$\mathbf{B} = \sum_{k=1}^{N} t^k z^k \mathbf{Q}^k, \tag{5.31}$$

$$\mathbf{D} = \sum_{k=1}^{N} t^k \left[\left(z^k\right)^2 + \frac{(t^k)^2}{12} \right] \mathbf{Q}^k, \tag{5.32}$$

where the mid-plane position z^k and thickness t^k of ply k are introduced as

$$z^k = \frac{h^k + h^{k-1}}{2} \quad \text{and} \quad t_k = h^k - h^{k-1} \tag{5.33}$$

and N represents the number of lamina.

Adolfsson and Gudmundson [9] related the strain increment produced by an array of cracks to the local crack face displacement.

The changes in stiffness matrices can be written as

$$\Delta\mathbf{A} = \sum_{k=1}^{N} \sum_{l=1}^{N} \sqrt{t^k \rho^k t^l \rho^l} \mathbf{C}_{EE}^{kl}, \tag{5.34}$$

$$\Delta\mathbf{B} = \sum_{k=1}^{N} \sum_{l=1}^{N} \sqrt{t^k \rho^k t^l \rho^l} \left(z^l \mathbf{C}_{EE}^{kl}\right), \tag{5.35}$$

$$\Delta\mathbf{D} = \sum_{k=1}^{N} \sum_{l=1}^{N} \sqrt{t^k \rho^k t^l \rho^l} \left(z^k z^l \mathbf{C}_{EE}^{kl} + \frac{t^k t^l}{4} \mathbf{C}_{BB}^{kl} \right), \tag{5.36}$$

where the matrix $\mathbf{C}$ takes account of the elastic properties and crack orientation relative to applied stress and can be written as

$$C_m^{kl} = \mathbf{Q}^k \left(\mathbf{N}^k\right)^T \boldsymbol{\beta}_m^{kl} \mathbf{N}^l \mathbf{Q}^l, \quad m = EE, BB, \tag{5.37}$$

where EE and BB denote pure extension and bending, respectively. The matrix $\mathbf{N}^k$ is defined from the constant unit normal vectors $\mathbf{n}^k$ for crack surfaces of ply k and can be written as

$$\mathbf{N}^k = \begin{bmatrix} \mathbf{n}_1^k & 0 & \mathbf{n}_2^k \\ 0 & \mathbf{n}_2^k & \mathbf{n}_1^k \end{bmatrix}. \tag{5.38}$$

Thus, damage matrices are proportional to the crack density and crack displacement vector $\boldsymbol{\beta}_m^{km}$. Adolfsson and Gudmundson [9] obtained a crack opening displacement matrix assuming that the different modes of crack opening displacements and tractions are independent and can be written as

$$\boldsymbol{\beta}_m^{kl} = \begin{bmatrix} \beta_{11(m)}^{kl} & 0 \\ 0 & \beta_{22(m)}^{kl} \end{bmatrix}, \quad m = EE, BB. \tag{5.39}$$

The assumption given above suggests that there will be no coupling between the crack opening displacements of different plies, hence

$$\boldsymbol{\beta}^{kl} = 0 \quad \text{for all } k \neq l. \tag{5.40}$$

The components of $\boldsymbol{\beta}^{kk}$ are derived using the relation between the stress intensity factors and energy release rate. The β_{11} components relate to crack face displacement in mode III anti-plane strain. β_{22} relates to mode I crack opening and can only be evaluated numerically by a series expression. The resulting components of the $\boldsymbol{\beta}^{kk}$-matrices are as given below:

$$\beta_{11(EE)}^{kk} = \frac{\pi}{2}\gamma_1^k \frac{8}{(\pi\rho^k)^2} \ln\left[\cosh\left(\frac{\pi\rho^k}{2}\right)\right], \tag{5.41}$$

$$\beta_{22(EE)}^{kk} = \frac{\pi}{2}\gamma_2^k \sum_{j=1}^{10} \frac{a_j}{(1+\rho^k)^j}, \tag{5.42}$$

for the components connected with pure extension (EE) and

$$\beta_{11(BB)}^{kk} = \frac{\pi}{16}\gamma_2^k \sum_{j=1}^{10} \frac{b_j}{(1+\rho^k)^j}, \tag{5.43}$$

$$\beta_{22(BB)}^{kk} = \frac{\pi}{16}\gamma_2^k \sum_{j=1}^{10} \frac{c_j}{(1+\rho^k)^j}, \tag{5.44}$$

for the components which must be added to take bending (BB) into account.

The quantities γ_1^k and γ_2^k are defined from the materials properties of ply k as

$$\gamma_1^k = \frac{1}{2G_{LT}^k}, \tag{5.45}$$

$$\gamma_2^k = \frac{1-\mu_{LT}^k\mu_{TL}^k}{E_T^k}. \tag{5.46}$$

The curve fit parameters a_j, b_j, and c_j are given in [9]. The results of the components connected to pure extension were obtained from Gudmundson and Zang [10], and the components required to be added to take bending into account are obtained from Adolfsson and Gudmundson [9] using the least square fit to the results from numerical integration.

After a certain crack density, the stiffness becomes saturated at a crack density known as the saturation crack density (ρ_0). However, as damage increases, matrix cracks may induce more severe damage at the tip of the crack such as debonding/delamination and fiber breakage. The debonding/delamination and fiber breakage damage models are developed in the next sections. In this mathematical model, it is assumed that the debonding/delamination will occur after matrix crack saturation and fiber breakage will start after debonding/delamination.

5.1.3.2 Debonding/Delamination

The effect of the debonding/delamination is modeled at the lamina mechanical property level. The effect of debonding/delamination induced by matrix cracks on the effective ply stiffness is estimated based on the continuum damage mechanics approach. Therefore, the degree of degradation of the mechanical properties in the ply due to matrix cracking-induced damage is associated with crack density and has the following form given by Shahid and Chang [11]:

$$E_{xx}^d(\rho) = E_{xx}(\rho), \tag{5.47}$$

$$E_{yy}^d(\rho) = E_{yy}(\rho)d_s, \tag{5.48}$$

$$\mu_{xy}^d(\rho) = \mu_{xy}(\rho), \tag{5.49}$$

$$\mu_{yx}^d(\rho) = \mu_{yx}(\rho)d_s, \tag{5.50}$$

$$G_{xy}^d(\rho) = G_{xy}(\rho)d_s, \tag{5.51}$$

where

$$d_s = e^{-[\bar{\varepsilon}_{yy}(\rho)/\bar{\varepsilon}_{yy}(\rho_0)]^\eta} \tag{5.52}$$

and ρ_0 is the crack saturation density of the ply, $\bar{\varepsilon}_{yy}(\rho_0)$ is the effective transverse strain of the ply at saturation crack density, and η is the shape parameter dictating the rate of degradation of the cracked ply due to matrix crack-induced damage. $E_{xx}(\rho)$, $E_{yy}(\rho)$, $G_{xy}(\rho)$, and $\mu_{xy}(\rho)$ are the effective mechanical properties including the effect of matrix cracking at crack density ρ. For graphite/epoxy composites, a value of η greater than or equal to 8 was found appropriate [11]. Here d_s is the material degradation factor due to matrix cracking-induced damage. Since η is constant for a given material, d_s varies with the effective strain ratio $(\bar{\varepsilon}_{yy}(\rho)/\bar{\varepsilon}_{yy}(\rho_0))$. In general, the model implies that matrix crack-induced debonding or delamination damage accumulates with an increase of the crack density.

Values of $E_{xx}(\rho)$, $E_{yy}(\rho)$, $G_{xy}(\rho)$, and $\mu_{xy}(\rho)$ are calculated using a genetic algorithm (GA) such that the numerical values of $\mathbf{A}^n$, $\mathbf{B}^n$, and $\mathbf{D}^n$ match with the $\mathbf{A}^{(c)}$, $\mathbf{B}^{(c)}$, and $\mathbf{D}^{(c)}$ matrices obtained from (5.26)–(5.28) for matrix crack density ρ_0. This ensures a smooth transition between matrix crack saturation and debonding/delamination. The optimization problem can be written in standard form as follows.

Minimize

$$\sqrt{\sum_{i=1}^{3}\sum_{j=1}^{3}[(\mathbf{A}_{ij}^n - \mathbf{A}_{ij}^{(c)})^2 + (\mathbf{B}_{ij}^n - \mathbf{B}_{ij}^{(c)})^2 + (\mathbf{D}_{ij}^n - \mathbf{D}_{ij}^{(c)})^2]}. \tag{5.53}$$

For design variables

$$0.5 \le \frac{E_{xx}(\rho)}{E_{xx}^0}; \quad \frac{E_{yy}(\rho)}{E_{yy}^0}; \quad \frac{G_{xy}(\rho)}{G_{xy}^0}; \quad \frac{\mu_{xy}(\rho)}{\mu_{xy}^0} \le 1.$$

Here $\mathbf{A}^n$, $\mathbf{B}^n$, and $\mathbf{D}^n$ are the stiffness matrices at the nth generation. E_{xx}^0, E_{yy}^0, G_{xy}^0, and μ_{xy}^0 are the mechanical properties of the virgin lamina. The GA [12] parameters crossover rate, mutation rate, and population size are selected to be 0.5, 0.01, and 20, respectively.

Using the mechanics properties, the ply stiffness due to the presence of debonding/delamination can be expressed as

$$Q_{xx}^M(\rho) = rE_{xx}^d(\rho), \tag{5.54}$$

$$Q_{yy}^M(\rho) = rE_{yy}^d(\rho), \tag{5.55}$$

$$Q_{yx}^{M}(\rho) = r\mu_{xy}^{d}(\rho)E_{xx}^{d}(\rho), \tag{5.56}$$

$$Q_{xy}^{M}(\rho) = r\mu_{yx}^{d}(\rho)E_{xx}^{d}(\rho), \tag{5.57}$$

$$Q_{ss}^{M}(\rho) = G_{xy}^{d}(\rho), \tag{5.58}$$

where

$$r = \left[1 - \mu_{xy}(\rho)\mu_{yx}(\rho)\right]^{-1}. \tag{5.59}$$

5.1.3.3 Fiber Breakage

The extreme damage mechanism in composites is fiber breakage because fibers are the primary load-carrying elements of fiber reinforced composites. Hence, fiber breakage is linked directly to the final failure of composites. For a unidirectional composite ply, sparse breaks of individual fibers may not cause the total failure of the composite. Based on the fiber bundle theory, the effect of the fiber breakage can be defined as

$$\begin{pmatrix} \bar{\sigma}_{xx} \\ \bar{\sigma}_{yy} \\ \bar{\sigma}_{xy} \end{pmatrix} = \begin{pmatrix} Q_{xx}^{M}(\rho) & Q_{xy}^{M}(\rho) & 0 \\ Q_{yx}^{M}(\rho) & Q_{yy}^{M}(\rho) & 0 \\ 0 & 0 & Q_{ss}^{M}(\rho) \end{pmatrix} \begin{pmatrix} d_f & 0 & 0 \\ 0 & d_f & 0 \\ 0 & 0 & d_f \end{pmatrix} \begin{pmatrix} \bar{\varepsilon}_{xx} \\ \bar{\varepsilon}_{yy} \\ \bar{\varepsilon}_{xy} \end{pmatrix}, \tag{5.60}$$

where d_f is the degradation coefficient for fiber breakage, which has the following form:

$$d_f = e^{-(A_f/\delta^2)^\beta} \tag{5.61}$$

and A_f is the extent of fiber failure area over which the stress is equal to or higher than the average tensile strength of the composite. Here δ is the fiber interaction length for the unidirectional composite under consideration, and β is associated with the rate of material degradation due to fiber breakage. Any value of β greater than or equal to 8 can produce the fiber breakage phenomenon [11]. As β is a constant for a given material, d_f varies with area ratio A_f/δ^2. For typical graphite/epoxy composites, the fiber interaction zone δ is very small and ranges between 0.002 and 0.007 inches and β is 8.

5.2 Data Reduction

For health monitoring of a composite helicopter rotor, blade response, root loads, and strains are candidate measurements. However, blade response, root loads, and strains are functions of the location of sensor, azimuth angle around the rotor disk ($0 \leq \psi \leq 2\pi$), and damage level. It is very difficult to handle these data for all sensor locations, azimuths, and damage level because of the large amount of data generated. Therefore, the time location of maximum values of these measurements or spatial locations at specific intervals are selected for obtaining data at various

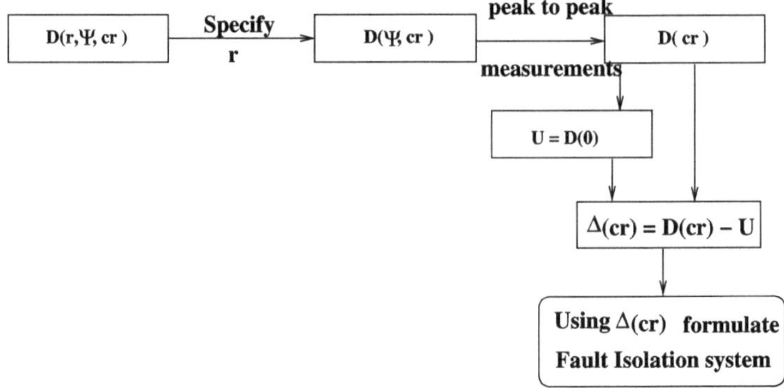

Fig. 5.3 Data reduction using fixed sensor locations and peak-to-peak sensor measurements

damage levels. For a specified sensor location, measurement will depend on azimuth (ψ) for a specified damage level. By taking the peak-to-peak value of the total signal obtained for ψ values between 0 and 360 degrees, the blade system response can be expressed as one scalar variable which is a function of the damage parameter only. Comparing the peak-to-peak value between the undamaged and damaged rotor blade results in a measurement delta which is an indicator of damage. The algorithm for data handling is shown in Fig. 5.3. **D** indicates the measurement for the damaged system, **U** indicates the measurement for the undamaged system, and cr represents the damage level.

This step of data reduction is important for the structural health monitoring (SHM) system. Sensors placed on a structure can quickly give very large amounts of data. Physical knowledge of the system and experience can be used to identify strategies for data reduction.

5.3 Life of the Structure

Besides detection and location of damage, one of the most desirable features of an SHM system is the ability to predict the remaining life of a structure. In this section, we develop a qualitative approach for predicting the remaining life in the composite helicopter rotor blade undergoing progressive damage accumulation. The stiffness degradation of the structure can be correlated to the life of the structure. The stiffness degradation is transformed as a function of life of the structure based on the phenomenological theory developed for fatigue life analysis of composite materials by Mao and Mahadevan [12]:

$$D = q\left(\frac{t}{t_f}\right)^{m_1} + (1-q)\left(\frac{t}{t_f}\right)^{m_2}, \tag{5.62}$$

Here $D = \frac{E_0 - E}{E_0 - E_f}$, E_0 is the initial stiffness ($t = 0$), E_f is the stiffness at final failure time t_f, and E is the stiffness at any instant of time t. We use the functional

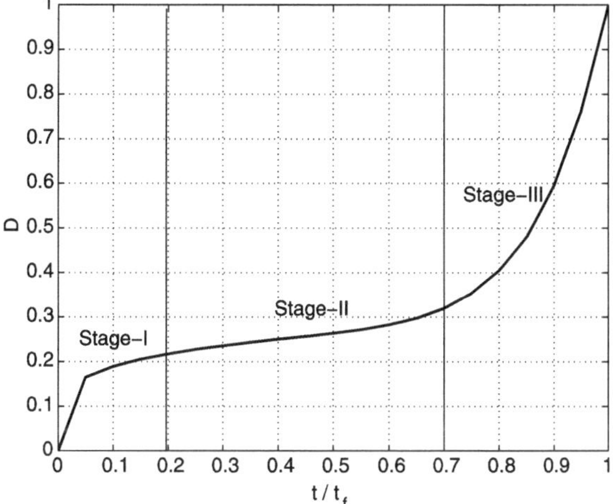

Fig. 5.4 Fatigue damage growth curve

relationship given in (5.62) as a curve fit to link the physics-based damage with the remaining life of the structure. Unlike other models for composite materials, the model in (5.62) captures the three phases of composite material degradation using one equation. Stage I is dominated by matrix cracking, Stage II by debonding/delamination, and Stage III by fiber breakage. Note that although (5.62) was originally developed to curve fit relating the continuum damage variable D with time, it can also be used for other variables which show a qualitatively similar behavior. Thus, (5.62) is a mathematical model of curves of the type shown in Fig. 5.4.

The values of the curve fitting parameters q, m_1, and m_2 are obtained by matching the stiffness reduction values obtained by physical damage modeling and by curve fitting (5.62) at the initial and final life of the structure and at the transition points of matrix cracking to debonding/delamination and debonding/delamination to fiber breakage.

5.4 Behavior of Composite Rotor Blade

To investigate the behavior of the composite rotor blade due to life consumption, a two-cell airfoil section beam with stiffness properties representing a stiff in-plane rotor is developed. The geometric properties and ply orientation of the two-cell airfoil section are shown in Fig. 5.5.

The ply elastic stiffness properties are $E_L = 30$ msi, $E_T = 3$ msi, $G_{LT} = 1.2$ msi, and $\mu_{LT} = 0.3$. The length (l) of the beam is 200 in. Since current rotor blades do not exploit composite couplings, the section is selected to have minimum couplings. Therefore, in the present study all the laminates are selected from the family of $(0/\pm 45/90)_s$ composites to minimize the coupling terms.

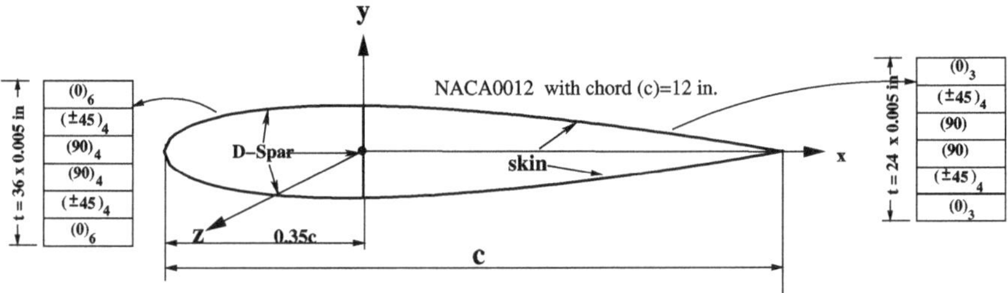

Fig. 5.5 Details of two-cell airfoil section beam

5.4.1 Effect on Cross-Sectional Stiffness

The stiffness reduction obtained due to the key damage modes such as matrix cracks, matrix crack-induced debonding/delamination, and fiber breakage is correlated to life consumption of the composite rotor blade. The stiffness reduction which is a function of the physical damage parameters is transformed to the stiffness reduction as a function of the life consumption of the composite rotor blade using (5.62).

Using the curve fit equation for bending stiffness gives

$$D_1 = 0.3\left(\frac{t}{t_f}\right)^{0.2} + 0.7\left(\frac{t}{t_f}\right)^{8}. \tag{5.63}$$

Here $D_1 = \frac{EI_{y0}-EI_y}{EI_{y0}-EI_{yf}}$, EI_{y0} is the initial bending stiffness ($t = 0$), EI_{yf} is the bending stiffness at final failure time t_f, and EI_y is the bending stiffness at any instant of time t. The same model fits the flap and lag bending stiffness since degradation affects the normalized flap and lag stiffness in a similar manner.

For torsion stiffness,

$$D_2 = 0.6\left(\frac{t}{t_f}\right)^{0.3} + 0.4\left(\frac{t}{t_f}\right)^{8}. \tag{5.64}$$

Here $D_2 = \frac{GJ_0-GJ}{GJ_0-GJ_f}$, GJ_0 is the initial torsion stiffness ($t = 0$), GJ_f is the torsion stiffness at final failure time t_f, and GJ is the torsion stiffness at any instant of time t. Here D_1 and D_2 can be interpreted as continuum damage variables in bending and torsion, respectively.

The bending and torsion stiffness reductions due to life consumption are shown in Figs. 5.6 and 5.7, respectively. To study the life consumption in various damage modes, the stiffness reduction plots are divided into three zones: matrix cracking, debonding/delamination, and fiber breakage. These correspond to Stages I, II, and III of the phenomenological model. From Figs. 5.6 and 5.7, it is observed that the life consumption in the matrix cracking zone is about 12–15% of the total life. The life consumption in the debonding/delamination zone is about 45–55% of the total life, and the remaining life of the structure is covered by the final failure fiber breakage. In these graphs, MC indicates matrix cracking, D/D indicates debonding/delamination, and FB represents fiber breakage.

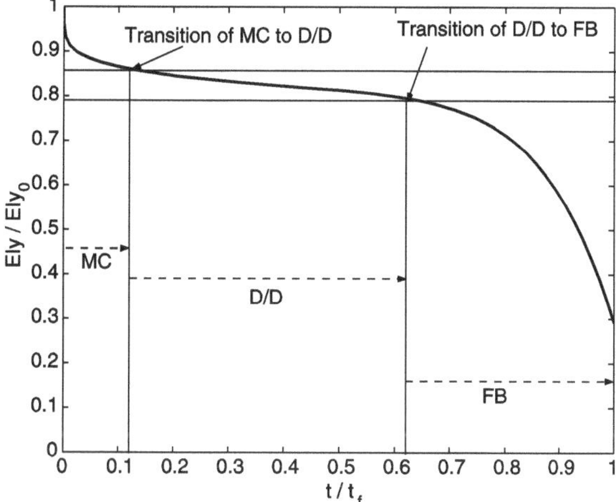

Fig. 5.6 Decrease in bending stiffness with increase in life consumption of the structure

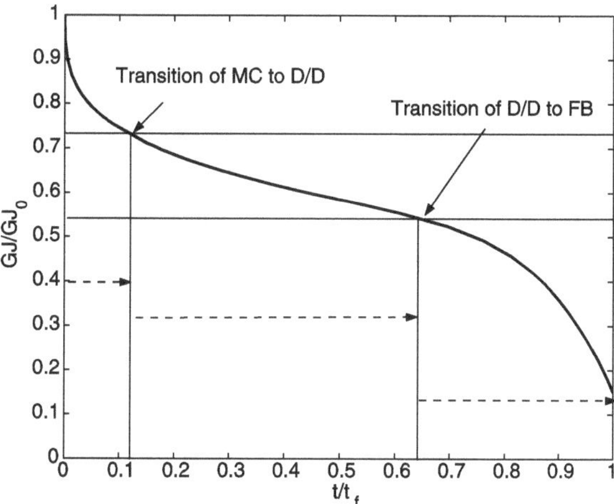

Fig. 5.7 Decrease in torsion stiffness with increase in life consumption of the structure

5.4.2 Effect on Static Response

Beam stiffness provides some indication of damage growth in composites. Changes in the beam stiffness can also change parameters such as the beam static response under a tip load or under gravity loads which can be measured at periodic intervals to assess the condition of the blade. The response of the airfoil section cantilever beam under unit static loads at the free end is obtained. Figures 5.8 and 5.9 show an increase in the normalized tip static responses ϕ_y (bending) and ϕ_z (torsion) of the beam with increasing life consumption. Here ϕ_y^0 and ϕ_z^0 are the tip static responses of the beams with virgin laminate. Similar to the stiffness plots, the plots of static responses are also divided into three damage zones. The threshold values of tip static deflection at the transition of these damage modes are useful for developing a ground-based maintenance and life management system. The maximum static

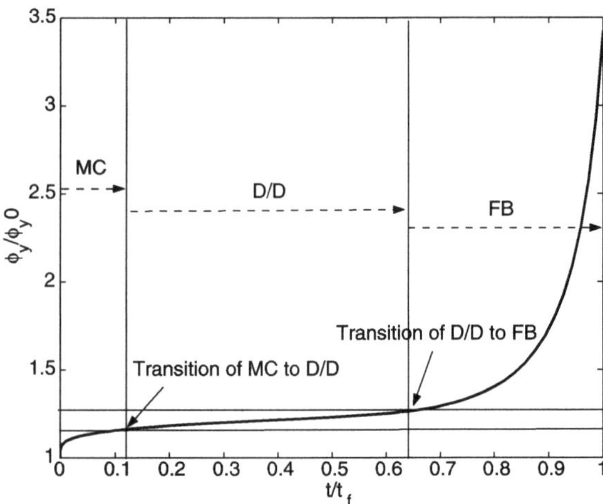

Fig. 5.8 Increase in tip bending response with increase in life consumption of the structure

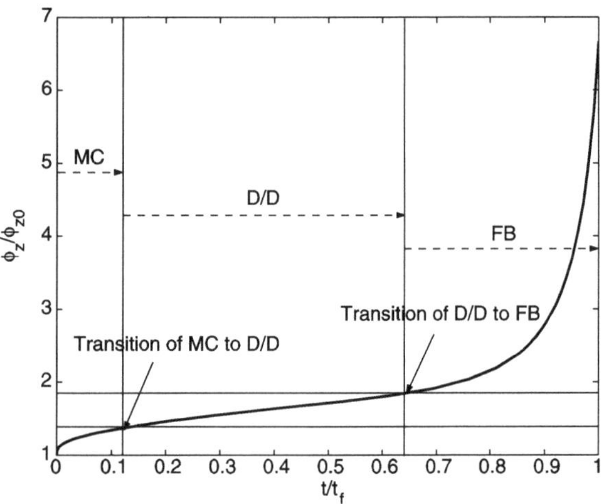

Fig. 5.9 Increase in tip torsion response with increase in life consumption of the structure

tip bending response after matrix crack saturation is about 1.15 times that of the virgin composite blade. The maximum static tip torsion response is about 1.5 times that of the virgin composite blade. Therefore, these maximum static values of tip static responses will give an indication that life of the composite beam in the matrix cracking damage zone is about 12–15% of the total life.

The rate of increase in static deflection in the D/D zone is very slow. However, the static deflection values at the transition of D/D and fiber breakage are very important because at this value the structure must be subjected to rigorous inspection and even replaced. Otherwise, fiber breakage may lead to catastrophic failure. The rate of variation of the tip static bending response in debonding/delamination damage mode is small compared to the torsion response. The maximum change in the tip bending slope due to debonding/delamination damage is about 1.3 times that of the virgin composite blade. By contrast, the maximum tip torsion response is about twice that of the virgin composite blade. When the tip bending slope exceeds 1.3 times that of the virgin beam and the tip torsion exceeds twice the virgin beam values, there is an indication that the composite beam has consumed 60–65% of the total life and entered the catastrophic damage mode. Therefore, the structure must be replaced when the tip bending slope exceeds 1.3 times that of the virgin beam and the tip torsion exceeds twice that of the virgin blade. We can see that mathematical models can allow the development of thresholds for condition monitoring.

Although fiber breakage is the catastrophic damage mode, the static responses show that the initial life of about 10–15% in this zone does not show a sudden increase in the static response. Therefore, the structure can be conservatively kept in working condition up to the transition point of D/D to fiber breakage. A more aggressive threshold at about 70–80% of the life consumption could be used to save costs at the cost of a higher risk of failure. In general, threshold selection in health monitoring involves a tradeoff between cost and risk.

5.5 Numerical Simulation of Measurement Deltas

For SHM, it is important to study the effects of damage on the behavior of the structure and the changes in measurable system properties due to damage. A four-bladed stiff in-plane hingeless composite rotor with progressive damage accumulation in the composite material is considered. The geometric properties and ply orientation of the two-cell airfoil section are the same as those given in the previous section. Results are obtained in forward flight at an advance ratio ($\mu = V/\Omega R$) of $\mu = 0.3$, a moderate thrust condition $C_T/\sigma = 0.07$, Lock number $\gamma = 6.34$, radius of gyration $\frac{mk_{m1}^2}{m_0 R^2} = 0.000174$, $\frac{mk_{m2}^2}{m_0 R^2} = 0.00061$, and $\frac{m}{m_0} = 1$. The effect of damage on global damage indicators such as blade tip response and blade root loads is first studied. Then, thresholds for these measurements are obtained using the matrix crack saturation point. Strains along the rotor blade are also evaluated as local damage indicators. As mentioned in previous chapters, a systematic study of the effect of damage growth on the structure is necessary to find useful damage indicators.

5.5.1 *Blade Tip Response*

The effect of matrix cracks in the whole blade on changes in the tip lag (ΔV), flap (ΔW), and torsion ($\Delta\phi$) response for varying crack densities are shown in Fig. 5.10. From Fig. 5.10, it is observed that the change in the tip lag bending deflection as compared to the tip flap bending deflection is negligible. It is also observed that the change in tip flap bending deflection reduces by 0.01 m in the initial phase of matrix cracking, and it increases steadily with an increase in crack density to 0.1 m. The change in tip torsion increases steadily up to 9 degrees, which is very high. Note

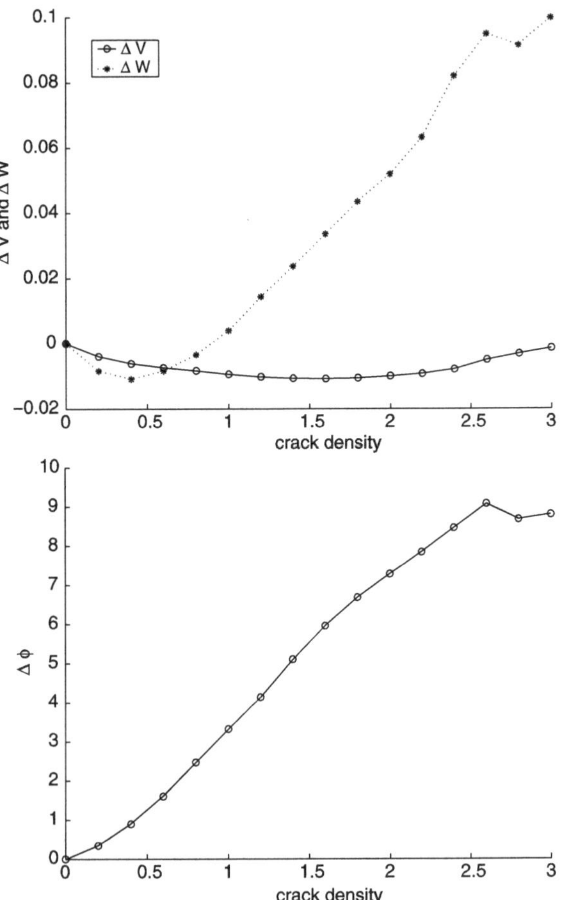

Fig. 5.10 Change in peak-to-peak tip lag (m), flap (m), and torsion (degrees) for increasing matrix crack density

that the values here are the change in peak-to-peak deformations across the azimuth between the damaged and undamaged blades. When matrix cracking effects start saturating, the change in tip torsion starts decreasing.

The effect of debonding/delamination in the whole blade on changes in the tip lag, flap, and torsion response for varying effective strain ratios are plotted in Fig. 5.11. During the transition from matrix cracking to debonding/delamination damage mode, the change in tip flap bending deflection value drops down to 0.07 m and the change in tip torsion value to 5 degrees. The change in tip flap bending and change in tip torsion remain nearly constant up to an effective strain of 0.88 and thereafter they increase up to 0.2 m and 12 degrees, respectively, due to an increase in the effective strain ratio.

The effect of fiber breakage in the whole blade on changes in the tip lag, flap, and torsion response for varying area ratios are plotted in Fig. 5.12. In this damage mode, the change in tip lag bending deflection is more affected and increases steadily up to 0.2 m. The change in tip flap bending deflection reduces from 0.17 m to 0.05 m

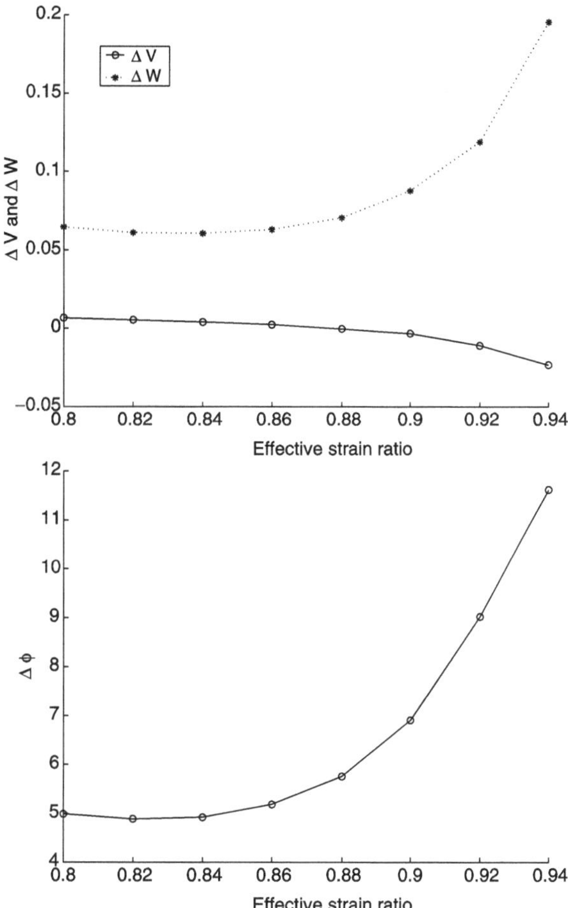

Fig. 5.11 Change in peak-to-peak tip lag (m), flap (m), and torsion (degrees) for increasing effective strain ratio due to debonding/delamination

and again increases up to 0.18 m. The change in tip torsion reduces from 15 degrees to 6 degrees.

The change in the system properties with damage evolution is not monotonic. This may be due to the substantial changes in the frequencies which pass each other as well as the rotor speed multiples as the damage grows. The aeroelastic effects of such frequency changes are considerable.

5.5.2 Blade Root Loads

The peak-to-peak change in blade root longitudinal (ΔF_x), lateral (ΔF_y), and vertical (ΔF_z) forces and rolling (ΔM_x), pitching (ΔM_y), and yawing (ΔM_z) moments, for varying crack densities, are plotted in Fig. 5.13. The effect of matrix cracking is more on the change in root vertical forces as compared to other forces. The change

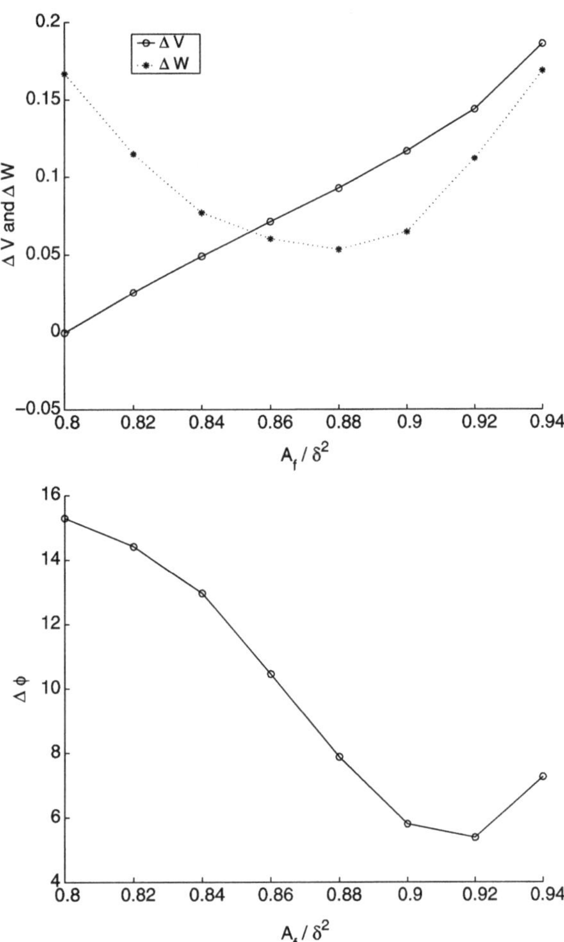

Fig. 5.12 Change in peak-to-peak tip lag (m), flap (m), and torsion (degrees) for increasing fiber area ratio due to fiber breakage

in force value of all forces reduces initially up to a crack density of 1 and then increases steadily. Curves of all the forces flatten toward the matrix crack saturation value. The maximum change in longitudinal and lateral blade root forces is about 1100 N and the maximum change in vertical blade root forces is about 2500 N. From the plots of peak-to-peak change in blade root bending moments, it is observed that the effect of the matrix cracking on blade root rolling and pitching moment is small compared to that on the blade root yawing moment. The change in root yawing moment decreases rapidly up to 4500 N-m for matrix crack density value 1, and thereafter it remains constant up to a crack density of 1.25, and again starts increasing.

The peak to peak changes in blade root forces and moments, for varying effective strain ratio, are plotted in Fig. 5.14. The changes in longitudinal and lateral forces remain almost constant in this damage mode. The value of the change in vertical root force remains almost constant up to an effective strain ratio of 0.88; thereafter,

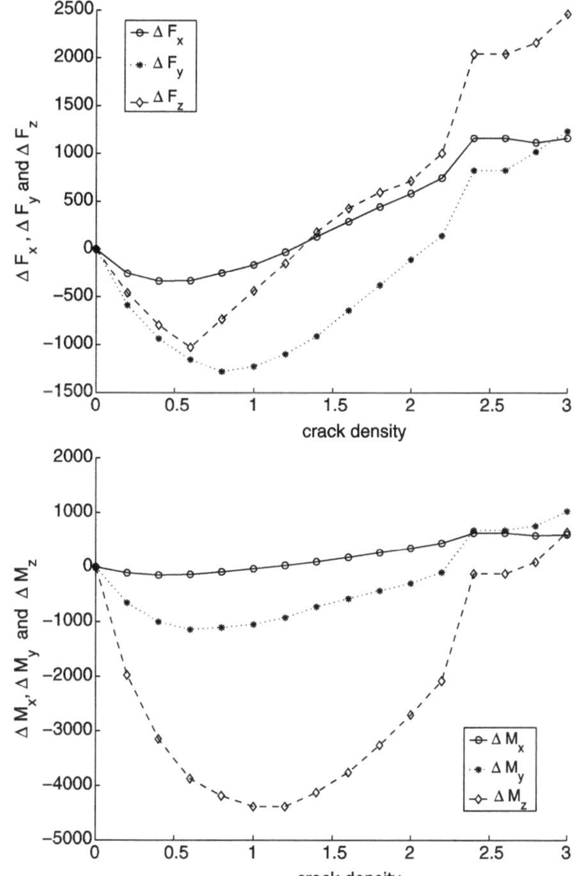

Fig. 5.13 Change in peak-to-peak blade root forces (N) and moments (N-m) for increasing matrix crack density

it starts increasing and reaches a maximum of about 9000 N for effective strain ratio 0.94.

From the plots of peak-to-peak change in blade root bending moments, it is observed that the effect of debonding/delamination on the change in pitching moments remains almost constant up to 0.88 and thereafter increases up to 4500 N-m. The change in yawing moment value decreases by about 1500 N-m, and the change in rolling moment value remains almost constant in this damage mode.

The peak-to-peak change in blade root forces and moments, for varying area ratio, are plotted in Fig. 5.15. The change in blade root longitudinal and lateral forces increases rapidly, and the change in root vertical force decreases rapidly.

From the plots of peak-to-peak change in blade root bending moments, it is observed that the effect of fiber breakage on the change in root rolling moment value is almost negligible. The change in root pitching moment decreases rapidly up to an area ratio of 0.88 and thereafter increases. The yawing moment remains almost constant up to an area ratio of 0.86 and thereafter increases rapidly up to 6000 N-m.

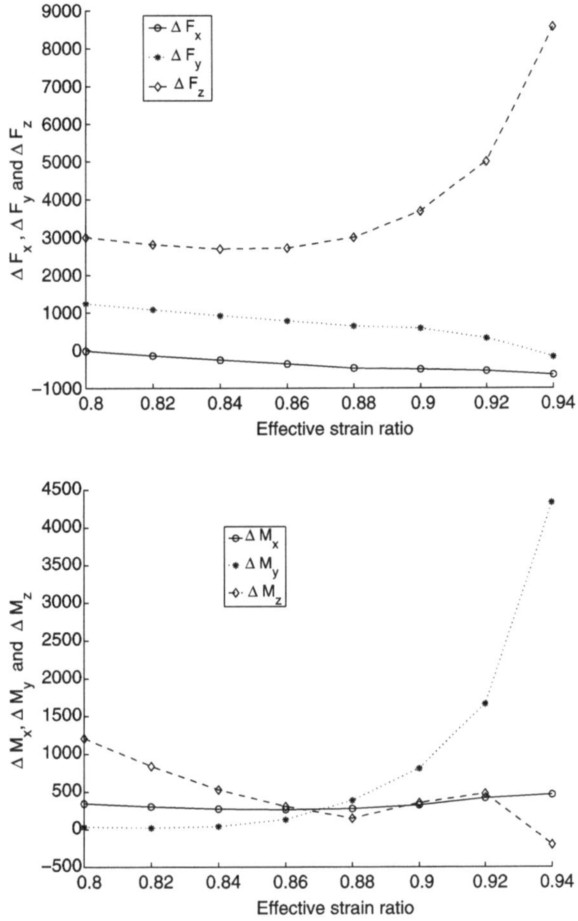

Fig. 5.14 Change in peak-to-peak blade root forces (N) and moments (N-m) for increasing effective strain ratio due to debonding/delamination

5.5.3 Thresholds Based on Matrix Crack Saturation

Since matrix crack saturation marks the start of more dangerous damage mechanisms, it is useful to use the change of rotor response at matrix cracking as a threshold for health monitoring. Table 5.1 shows the values of this blade response as loads at matrix crack saturation and the maximum change in these values from the baseline undamaged blade. The maximum changes in these values can be used as a threshold to flag the rotor blade condition as prone to damage. Such a blade could then be placed on a "watch list" subject to nondestructive testing techniques periodically.

5.5.4 Strains

Strains are useful local indicators of damage. We consider damage at five locations on the blade ranging from the root to the tip. To investigate the implementation

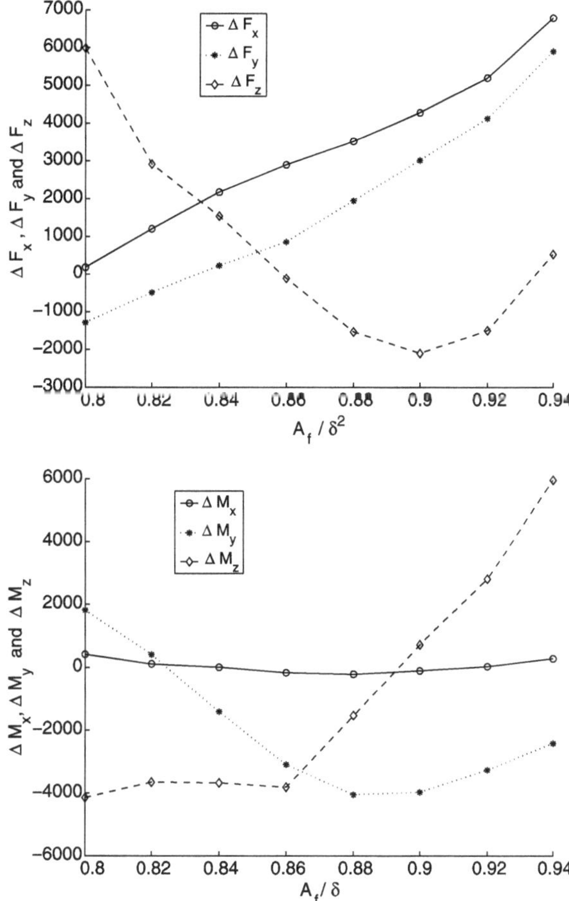

Fig. 5.15 Change in peak-to-peak blade root forces (N) and moments (N-m) for increasing fiber area ratio due to fiber breakage

Table 5.1 Change in peak-to-peak value of tip lag (m), tip flap (m), tip torsion (deg), root forces (N), and root moments (N-m) at matrix crack saturation and maximum change during matrix crack damage

	ΔV	ΔW	$\Delta\phi$	ΔF_x	ΔF_y	ΔF_z	ΔM_x	ΔM_y	ΔM_z
At Saturation	−0.0015	0.1	8.8	1.16E3	1.24E3	2.46E3	0.60E3	1.03E3	0.65E3
Max. Change	−0.0109	0.1	9.1	1.16E3	−1.28E3	2.46E3	0.63E3	−1.14E3	−4.38E3

of strain measuring instrumentation on the rotor blade, the peak-to-peak change in shear strains is calculated at various locations for matrix cracking, debonding/delamination, and fiber breakage. The results are shown in Figs. 5.16, 5.17, and 5.18, respectively. All the strains are calculated on the top side of the beam and along the line passing through the point 0.35c on the two-cell airfoil section. From these plots, it is observed that the maximum change in strains among all five locations is at the location where the damage is situated. The change in strains in-

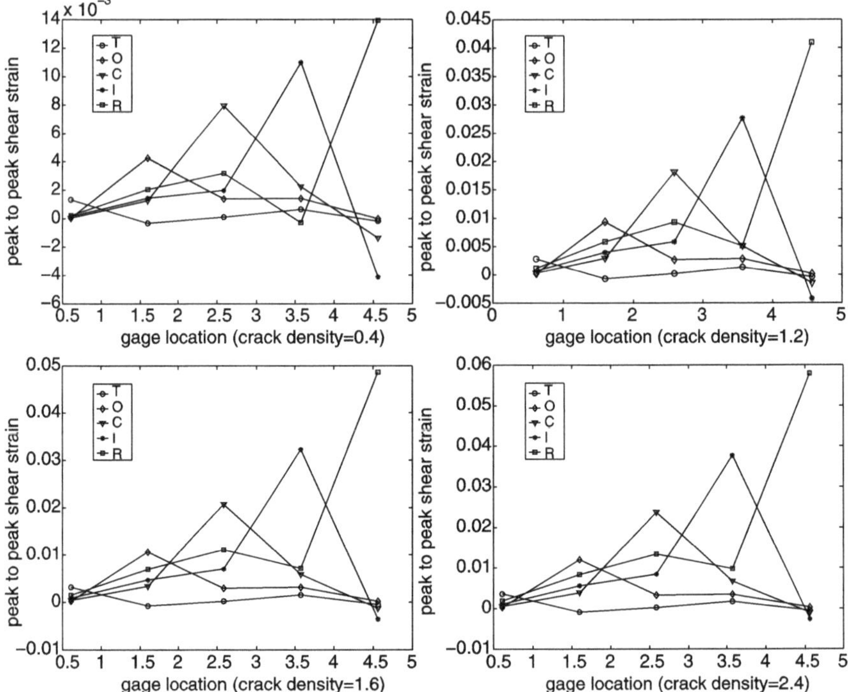

Fig. 5.16 Change in peak-to-peak shear strain measured at various locations for various crack densities and cracks at various locations (tip = 0 m and root = 5 m). Crack locations: T = Tip, O = Outboard, C = Center, I = Inboard, R = Root

creases with increasing damage level. From the plots of all three modes of damages, it is observed that the shear strain can be useful to isolate damage locations and damage level.

5.6 Predicting Life Consumption

Until now, the results have focused on static variables such as stiffness and blade tip deflection under a static loading. Obviously, there is a close relationship between stiffness and static displacements of the structure. However, static deflection cannot be used for online health monitoring. To simulate online measurements, an aeroelastic analysis of a composite rotor blade is performed. The composite damage models are integrated into the rotor aeroelastic analysis. The rotor system behavior is further linked with life consumption through the relationship between stiffness and time. The measurement deltas between a damaged and undamaged blade are used to develop two genetic fuzzy systems (GFSs) for the prediction of global and local physical damage and life consumption. The development of these GFSs for SHM is discussed next.

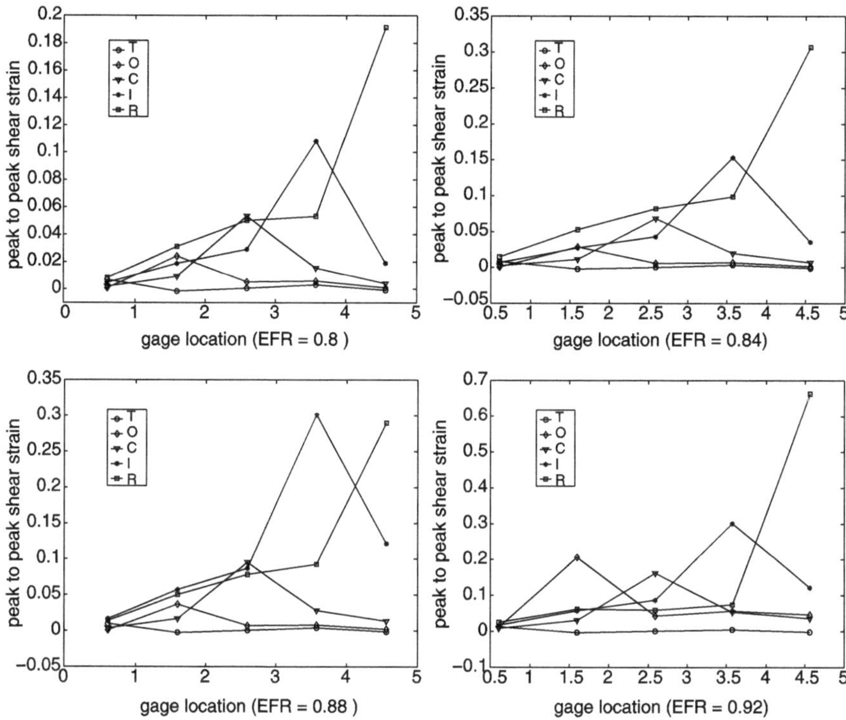

Fig. 5.17 Change in peak-to-peak shear strain measured at various locations for various effective strain ratios and debonding/delamination at various locations (tip = 0 m and root = 5 m). D/D locations: T = Tip, O = Outboard, C = Center, I = Inboard, R = Root

5.6.1 Development of Genetic Fuzzy System

There are two possibilities of damage in composite rotor blades. The first possibility is that the damage will be approximately uniformly distributed along the whole blade (global damage detection). The second possibility is that the damage will be localized. In this section, we formulate two GFSs. The first GFS is for predicting the physical damage parameter and life consumption parameter in the matrix cracking and debonding/delamination zones along the whole blade. The second GFS is for predicting the physical damage parameter and life consumption parameter in the matrix cracking and debonding/delamination zones in various parts of the blade (local damage detection). The fiber breakage damage mode in composite materials is considered as a catastrophic damage mode, and detection of such a damage zone is not useful. Therefore, the fiber breakage zone is not considered while designing the GFS. The schematic diagram of the SHM system development process is shown in Fig. 5.19. The difference between the global and local SHM lies in the choice of sensor measurements $\mathbf{z}$ in Fig. 5.19, as discussed next.

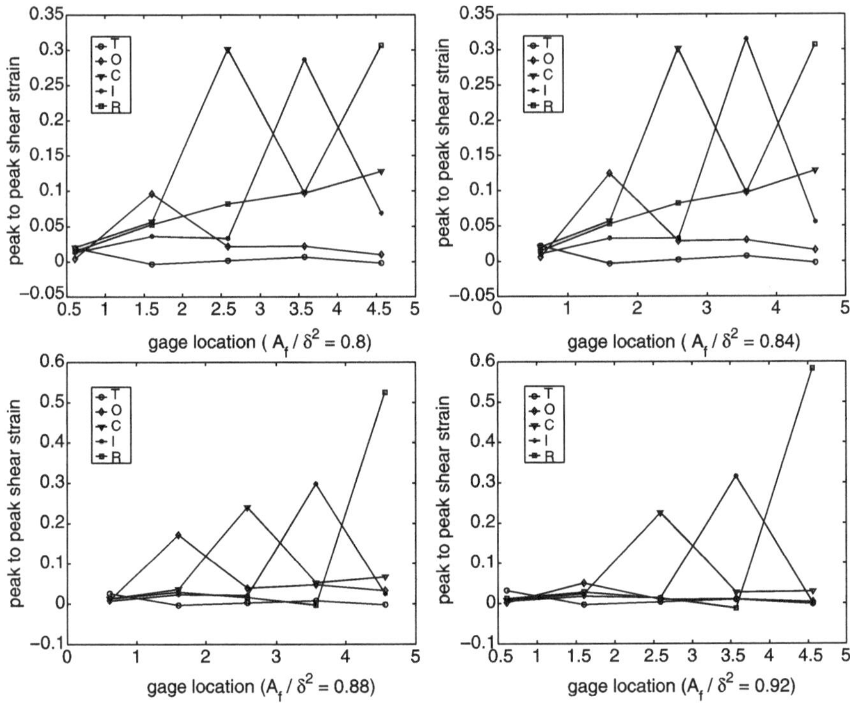

Fig. 5.18 Change in peak-to-peak shear strain measured at various locations for various area ratios and fiber breakage at various locations (tip = 0 m and root = 5 m). Fiber breakage locations: T = Tip, O = Outboard, C = Center, I = Inboard, R = Root

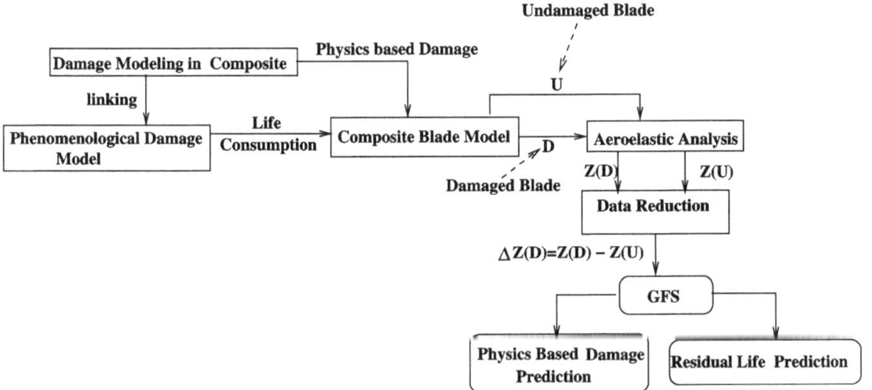

Fig. 5.19 Schematic representation of development of SHM

5.6.1.1 Global Damage Detection

The GFS for global damage detection is designed using the displacement-, force-, and moment-based measurement deltas. The objective is to find the mapping between measurement deltas and various damage levels.

In this fuzzy system, physical damage parameters and life consumption parameters in the matrix cracking and debonding/delamination zones are expressed as words. To get a degree of resolution of the extent of physical damage and life consumption, the physical damage parameter and life consumption parameter are allowed to have several zones based on the physical damage and split into linguistic variables. These classifications are based on the numerical results obtained for matrix cracking and debonding/delamination earlier in this chapter. We first consider the displacement measurement deltas for defining the rules, which show approximately monotonic behavior with respect to matrix cracking and debonding/delamination (Figs. 5.10 and 5.11). Next, the force and moment measurement deltas are considered for prediction of matrix cracking and debonding/delamination (Figs. 5.13 and 5.14). The first step in fuzzy logic is to transform typical measurements into fuzzy linguistic measures. This is done using the plots of change in peak-to-peak displacements and matrix crack density shown in Fig. 5.10 and the plots of change in peak-to-peak displacements and debonding/delamination shown in Fig. 5.11. Table 5.2 shows the linguistic and numerical measures based on the displacement measurement deltas. This table also shows the relation between the physical damage parameters and the life consumption parameters of the composite rotor blade. Based on the experience of the prognostic actions for composite structures, three prognostic actions for the three stages of the life of the composite blade are shown in this table.

The measurement deltas ΔW, $\Delta\phi$, ΔF_x, ΔF_y, ΔF_z, ΔM_x, ΔM_y, and ΔM_z are treated as fuzzy variables. Fuzzy sets with Gaussian membership functions are used to define these input variables. Changes in the measurement deltas are calculated using the aeroelastic analysis for different levels of damages along the whole blade, as explained in Sect. 5.5. The midpoints of the Gaussian function are calculated by normalizing the changes in measurement deltas with its maximum values. The genetic fuzzy system is tested using the normalized noisy measurement delta (x) which is derived from the noise model given in (2.6). The addition of noise to the simulations is needed to make it realistic and develop a robust model-based diagnostic system.

Rules for the fuzzy system are obtained by fuzzification of the numerical values obtained from an aeroelastic analysis of the composite helicopter blade in forward flight. The fuzzy sets corresponding to ΔW, $\Delta\phi$, ΔF_x, ΔF_y, ΔF_z, ΔM_x, ΔM_y, and ΔM_z are generated by taking the change in measurements obtained from the aeroelastic analysis solution as midpoints of membership functions corresponding to a damage level. This strategy for selecting the midpoint ensures that the maximum degree of membership ($\mu = 1$) for each fuzzy set occurs at the values of m since the Gaussian function is the highest at the midpoint. The standard deviations of the Gaussian membership functions are calculated using a GA for maximization of the success rate, as discussed in earlier chapters.

Table 5.2 Linguistic classification of damage for the GFS for global damage detection

No.	Physics-based rule	Damage level	Life consumption (Residual life)	Prognostic action
1	Undamaged	Nil crack density	Nil (100%)	**OK**
2	V. Small CD	Crack density 0–0.8	0–2.5% (97.5%)	**OK**
3	Small CD	Crack density 0.8–1.2	2.5–5% (95%)	**OK**
4	Considerable CD	Crack density 1.2–1.6	5–7% (93%)	**OK**
5	High CD	Crack density 1.6–2.0	7–8.5% (91.5%)	**OK**
6	Very high CD	Crack density 2.0–2.4	8.5–10% (90%)	**OK**
7	Saturation CD	Crack density 2.4–3.0	10–12 % (88%)	**OK**
8	Transition of MC to D/D	CD 3.0[a] to ESR 0.8[b]	12–20% (80%)	**WATCH**
9	Slight D/D	ESR 0.8–0.88	20–43% (67%)	**WATCH**
10	Moderate D/D	ESR 0.88–0.9	43–50% (50%)	**WATCH**
11	Severe D/D	ESR 0.9–0.92	50–56% (44%)	**WATCH**
12	Extreme D/D	ESR 0.92–0.94	56–62% (38%)	**REMOVE**

[a]Saturation crack density (CD)

[b]The ESR from where the effects of D/D become considerable

Prognostic actions: **OK**: Blade is OK, no action is required. **WATCH**: Put blade under watch. **REMOVE**: Remove blade. Take for thorough inspection

5.6.1.2 Local Damage Detection

The GFS for local damage detection predicts the physical damage parameter and life consumption parameter in the matrix crack and debonding/delamination zones and finds the location of the damage along the blade.

Inputs to the GFS are strain-based measurement deltas at five locations, and outputs are physical damage parameters and the life consumption parameter at different locations. The objective is to find a mapping between the measurement deltas and physical damage parameters and life consumption parameters at five different locations. In this fuzzy system, the location parameters are crisp numbers. The damage locations are divided and labeled as discussed in Fig. 3.2.

To get a degree of resolution of the extent of physical damage and life consumption, physical damage parameters and life consumption parameters at each location are allowed several levels and split into linguistic variables. Fuzzy logic rules are defined based on the shear strains obtained for a few key physical damage parameters. These shear strains are first obtained for the physical damage parameters and then linked with the life of the blade. The linguistic relations of the key life consumption parameters and physical damage parameters are shown in Table 5.3. Similar to the global SHM system, the prognostic actions for the three stages of the life of the composite structure are shown in the table.

Strain-based measurement deltas $\Delta\varepsilon_{\mathrm{Tip}}$, $\Delta\varepsilon_{\mathrm{Outboard}}$, $\Delta\varepsilon_{\mathrm{Center}}$, $\Delta\varepsilon_{\mathrm{Inboard}}$, and $\Delta\varepsilon_{\mathrm{Root}}$ are treated as fuzzy variables. Fuzzy sets with Gaussian membership functions are used to define these input variables. Changes in strains (measurement

Table 5.3 Linguistic classification of damage for local damage detection

Damage name	Damage level	Life consumption (Residual life)	Prognostic action
Undamaged	Matrix crack density zero	Nil (100%)	**OK**
Small CD	Matrix crack density 0.4	About 2% (98%)	**OK**
Moderate CD	Matrix crack density 1.2	About 5% (95%)	**OK**
High CD	Matrix crack density 1.6	About 7% (93%)	**OK**
Very high CD	Matrix crack density 2.4	About 10% (90%)	**OK**
Slight D/D	ESR 0.8	About 20% (80%)	**WATCH**
Moderate D/D	ESR 0.88	About 43% (67%)	**WATCH**
Severe D/D	ESR 0.92	About 56% (44%)	**REMOVE**

Prognostic actions: **OK**: Blade is OK, no action is required. **WATCH**: Put blade under watch. **REMOVE**: Remove blade. Take for thorough inspection

deltas) are calculated using an aeroelastic analysis for a combination of five different locations and seven different levels of damages and are shown in Figs. 5.16 and 5.17. Formulation of the GFS and calculation of the success rate are performed using the algorithm discussed in previous chapters.

5.6.2 *Testing of Genetic Fuzzy System*

The global GFS and the local GFS are tested at various noise levels. All the measurements are normalized with their maximum value. The details of the tests are discussed next.

5.6.2.1 Global Damage Detection

Two key issues are discussed in the next two subsections using numerical simulations. The first is the use of different sets of measurement deltas, and the second is the effect of training noise of the GFS.

Effect of Different Measurement Delta Sets The midpoints and standard deviations of the Gaussian membership functions for the rules of prediction of physical damage parameters and life consumption parameters along the whole blade with a training noise level of 0.15 are given in Table 5.4. Success rates of the GFS are calculated for various noise levels of 0.03, 0.05, 0.10, 0.15, and 0.20. The GFS is tested using four sets of measurement deltas: (1) tip displacement only, (2) blade root forces only, (3) blade root moments only, and (4) all three. The performance of the SHM for the four sets of measurement deltas gives an idea about the reliability of the SHM based on different measurement instruments. It may be possible to implement the displacement-based SHM in an old helicopter to achieve economy and simplicity. The root load-based SHM can be implemented when the root loads can be

Table 5.4 Midpoints and standard deviations of rules for the GFS for global damage detection

No.	ΔW	$\Delta\theta$	ΔF_x	ΔF_y	ΔF_z	ΔM_x	ΔM_y	ΔM_z
1	0.00 (0.10)	0.00 (0.12)	0.00 (0.19)	0.00 (0.16)	0.00 (0.14)	0.00 (0.18)	0.00 (0.12)	0.00 (0.19)
2	−0.07 (0.11)	0.09 (0.13)	−0.11 (0.13)	−0.52 (0.17)	−0.05 (0.14)	−0.07 (0.16)	−0.18 (0.19)	−0.49 (0.13)
3	0.03 (0.17)	0.32 (0.11)	−0.12 (0.15)	−0.96 (0.18)	−0.06 (0.17)	−0.04 (0.17)	−0.34 (0.19)	−1.00 (0.17)
4	0.15 (0.11)	0.49 (0.11)	0.11 (0.12)	−0.70 (0.16)	0.02 (0.13)	0.18 (0.11)	−0.25 (0.15)	−0.95 (0.13)
5	0.27 (0.17)	0.64 (0.17)	0.38 (0.16)	−0.30 (0.15)	0.08 (0.13)	0.43 (0.13)	−0.14 (0.20)	−0.75 (0.12)
6	0.42 (0.10)	0.76 (0.15)	0.75 (0.13)	0.29 (0.15)	0.20 (0.18)	0.80 (0.19)	0.07 (0.15)	−0.33 (0.15)
7	0.57 (0.14)	0.84 (0.17)	1.00 (0.19)	0.83 (0.19)	0.33 (0.14)	1.00 (0.18)	0.29 (0.12)	0.06 (0.14)
8	0.51 (0.16)	0.67 (0.17)	0.49 (0.13)	1.00 (0.16)	0.40 (0.13)	0.77 (0.19)	0.18 (0.13)	0.22 (0.15)
9	0.42 (0.17)	0.52 (0.15)	−0.21 (0.14)	0.76 (0.11)	0.44 (0.16)	0.50 (0.13)	0.07 (0.11)	0.16 (0.18)
10	0.50 (0.15)	0.61 (0.18)	−0.42 (0.16)	0.50 (0.18)	0.49 (0.16)	0.49 (0.13)	0.20 (0.18)	0.06 (0.19)
11	0.65 (0.14)	0.77 (0.11)	−0.45 (0.16)	0.37 (0.18)	0.64 (0.16)	0.60 (0.16)	0.41 (0.20)	0.10 (0.18)
12	0.98 (0.19)	1.00 (0.12)	−0.51 (0.10)	0.07 (0.18)	1.00 (0.11)	0.71 (0.20)	1.00 (0.12)	0.03 (0.17)

obtained from the existing instrumental arrangements and the SHM based on all the measurement deltas can be implemented when more accurate prediction is needed. Table 5.5 shows the success rates for all the rules for four different sets of measurement deltas at various noise levels; these results are summarized in Fig. 5.20. From Table 5.5 and Fig. 5.20, it is observed that the GFS with displacement-based measurement deltas gives good results at the noise level of 0.05 with an average success rate of 95.55% and a minimum success rate of 80.90% (transition of matrix cracking to debonding/delamination) and starts deteriorating rapidly with a further increase in noise level. The critical performance of the SHM system can be judged based on the success rate of the key fuzzy rules. The key fuzzy rules for the current SHM are rule number 7 and rule number 12. Rule number 7 gives an indication of the beginning of a severe damage mode, debonding/delamination, and changing to new prognostic action "Put blade under watch." Rule number 12 gives an indication about the beginning of a catastrophic damage, fiber breakage. The SHM performs satisfactorily for the two key rules by giving success rates of 98% and 100% for rule number 7 and 12, respectively, at a noise level of 0.05.

The GFS with force-based measurement deltas gives a success rate of 100% at the noise level of 0.05. For a noise level of 0.1, the GFS with force-based measurement deltas gives an average success rate of 99% and a minimum success rate of 91.3% (Severe D/D) and starts falling for higher noise levels of 0.15 and 0.20 by giving average success rates of 96% and 89.8% and minimum success rates of 76% and 63.90%, respectively. The GFS with force-based measurement deltas gives a 100% success rate for both the key rules even at a higher noise level of 0.15.

The GFS with moment-based measurement deltas gives a success rate of 100% at the noise level of 0.05. For a noise level of 0.1, the GFS with moment-based measurement deltas gives an average success rate of 98.05% and a minimum success rate of 91.4% (Slight D/D) and starts falling for higher noise levels of 0.15 and 0.20 by giving average success rates of 96% and 89.8%, respectively, and minimum success rates of 76% and 63.90%, respectively. The GFS with moment-based measurement deltas also gives good performance for the key rules by giving a success rate of about 96% for rule 7 and a success rate of about 100% for rule 12, for a higher noise level of 0.15.

When all measurement deltas are considered for matrix cracking and debonding/delamination detection, the GFS gives a success rate of 100% percent up to a noise level of 0.10 and also gives good results with higher noise levels of 0.15 and 0.20 by giving average success rates of 99.21% and 97.02%, respectively, and minimum success rates of 95.30% and 86.30%, respectively. Thus, the use of both displacement- and load-based measurements results in a health monitoring system with high levels of accuracy even with noisy data.

It is noted that the GFS with displacement-based, force-based, and moment-based measurement deltas gives a success rate of about 100% for rule 12, which indicates the beginning of catastrophic damage. Therefore, the GFS can be used to give a warning about the beginning of catastrophic damage with a minimum number of measurement deltas in a highly noisy environment. It can also be noted that, though the noise tolerance capacity of the GFS is dependent on the type of measurement delta, the accuracy of the measurement instrument will also have an influence

Table 5.5 Success rates for various testing noise levels and training noise level of 0.15 for global damage detection using displacement, force, moments, and all measurements

No.	Displacements				Forces				Moments				All			
	$S_{R0.05}$	$S_{R0.10}$	$S_{R0.15}$	$S_{R0.20}$	$S_{R0.05}$	$S_{R0.10}$	$S_{R0.15}$	$S_{R0.20}$	$S_{R0.05}$	$S_{R0.10}$	$S_{R0.15}$	$S_{R0.20}$	$S_{R0.05}$	$S_{R0.10}$	$S_{R0.15}$	$S_{R0.20}$
1	96.90	79.70	69.90	60.20	100.00	100.00	99.60	95.50	100.00	100.00	100.00	97.50	100.00	100.00	100.00	100.00
2	97.80	81.60	68.70	50.50	100.00	100.00	97.00	90.30	100.00	100.00	100.00	96.40	100.00	100.00	100.00	99.90
3	100.00	97.50	77.60	65.10	100.00	100.00	99.70	95.90	100.00	99.90	94.70	90.60	100.00	100.00	100.00	99.30
4	100.00	81.20	46.10	29.10	100.00	100.00	89.30	66.30	100.00	94.90	74.30	45.80	100.00	100.00	98.30	90.20
5	100.00	92.40	69.40	49.90	100.00	100.00	100.00	100.00	100.00	100.00	99.90	97.80	100.00	100.00	100.00	100.00
6	98.10	61.40	35.50	24.40	100.00	100.00	100.00	99.80	100.00	100.00	99.50	89.60	100.00	100.00	100.00	100.00
7	97.70	73.60	55.10	41.70	100.00	100.00	100.00	100.00	100.00	100.00	95.50	89.00	100.00	100.00	100.00	100.00
8	80.90	48.80	27.80	21.10	100.00	100.00	100.00	100.00	100.00	99.80	87.00	68.00	100.00	100.00	100.00	99.50
9	98.40	73.90	60.40	48.20	100.00	100.00	96.40	91.30	100.00	91.40	75.40	60.80	100.00	100.00	95.30	86.30
10	82.30	46.60	31.60	23.70	100.00	97.20	88.30	78.00	100.00	93.20	76.70	65.20	100.00	100.00	98.90	94.40
11	94.50	66.00	46.90	34.00	100.00	91.30	76.30	63.90	100.00	97.40	79.70	67.70	100.00	100.00	98.00	94.60
12	100.00	100.00	100.00	96.60	100.00	100.00	100.00	96.60	100.00	100.00	100.00	100.00	100.00	100.00	100.00	100.00
Avg.	95.55	75.23	57.42	45.38	100.00	99.04	95.55	89.80	100.00	98.05	90.23	80.70	100.00	100.00	99.21	97.02
Min	80.90	46.60	27.80	21.10	100.00	91.30	76.30	63.90	100.00	91.40	74.30	45.80	100.00	100.00	95.30	86.30

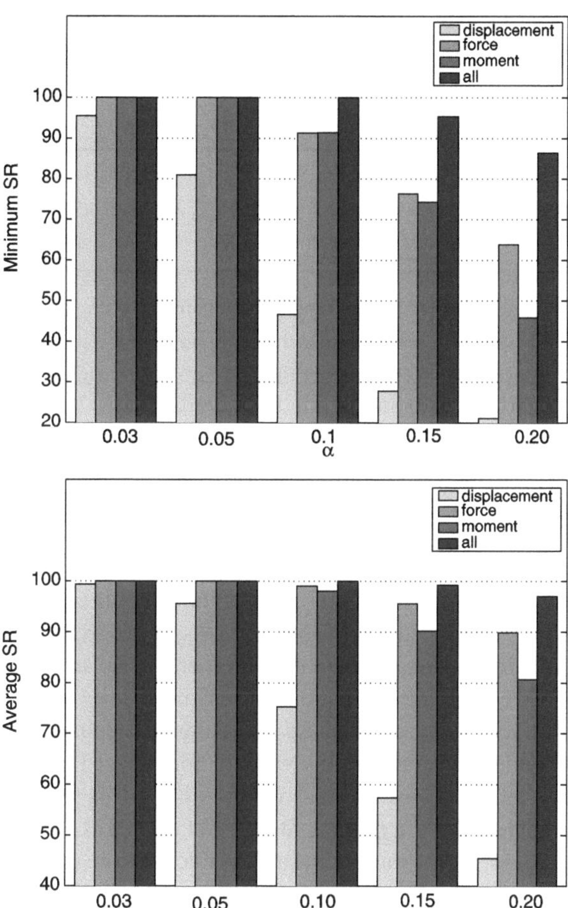

Fig. 5.20 Effect of various measurements on minimum and average success rate of genetic fuzzy logic system for global damage detection

on the accuracy of damage prediction. For example, the displacements are measured directly using optical blade tracking, and the laser Doppler vibrometer (LDV) gives more accurate measurements than the force and moment measurements, which are estimated using vibration level measurements.

Effect of Training Noise The previous results used a training noise level of 0.15. The effect of different levels of noise in the training data is now considered. The GFS is trained using noisy data with noise levels (training noise level) of 0.05, 0.10, and 0.15. These GFSs are then tested using noisy data for noise levels of 0.05, 0.10, and 0.15. The effects of various training noise levels are studied by comparing the minimum and average success rates of the GFS with training noise levels of 0.05 and 0.10 with the minimum and average success rates of the GFS with a training noise level of 0.15 for various testing noise levels. $\Delta S_{R0.05}$ indicates the difference between the success rates with a training noise level of 0.05 compared with the

Table 5.6 Effect of training noise level on minimum success rate at various noise levels for global damage detection using displacement, force, moments, and all measurements

α	Displacements		Forces		Moments		All	
	$\Delta S_{R0.05}$	$\Delta S_{R0.10}$	$\Delta S_{R0.05}$	$\Delta S_{R0.10}$	$\Delta S_{R0.05}$	$\Delta S_{R0.10}$	$\Delta S_{R0.05}$	$\Delta S_{R0.10}$
0.05	15.60	8.10	0	0.02	0	0	0	0
0.10	16.10	11.40	−2.20	4.80	4.90	1.70	0	0.02
0.15	13.30	6.90	−3.20	14.20	1.90	4.80	−2.70	0

$\Delta S_{R0.05}$ = Difference between the minimum SR for training at $\alpha = 0.15$ and $\alpha = 0.05$
$\Delta S_{R0.10}$ = Difference between the minimum SR for training at $\alpha = 0.15$ and $\alpha = 0.10$
Negative value indicates better SR than the SR for training at $\alpha = 0.15$

success rates with a training noise level of 0.15. A similar definition is used for $\Delta S_{R0.10}$.

The effects of training noise on the minimum success rates for different measurement data sets at various (testing) noise levels are shown in Table 5.6. The minimum success rate gives an indication of the reliability of the GFS. From Table 5.6 it is observed that the highest values of $\Delta S_{R0.05}$ and $\Delta S_{R0.10}$ for the minimum success rate with displacement-based measurement deltas are about 16.1 and 11.4, respectively. This result indicates that the training noise level of 0.15 is a good option for the GFS with the displacement-based measurement delta. The highest value of $\Delta S_{R0.05}$ for the minimum success rate with the force-based measurement deltas is about −3.20 and the highest value of $\Delta S_{R0.10}$ is 14.2. This shows that for the force-based measurement delta, a training noise level of 0.05 is a good option. The highest value of $\Delta S_{R0.05}$ for the minimum success rate with the moment-based measurement delta is about 4.90 and the highest value of $\Delta S_{R0.10}$ is 4.80. This shows that for the moment-based measurement delta, a training noise level of 0.15 is a good option. The highest value of $\Delta S_{R0.05}$ for the minimum success rate with all measurement deltas together is about −2.70 and the highest value of $\Delta S_{R0.10}$ is 0.02. This shows that for all the measurement deltas, a training noise level of 0.05 is a good option.

The effects of training noise on the average success rates for different measurement sets at various (testing) noise levels is shown in Table 5.7. The average success rate gives an indication of the overall performance of the system. From Table 5.7, it is observed that the GFS with training noise level of 0.15 gives a good success rate except for some cases with a training noise level of 0.05. The highest negative value $\Delta S_{R0.05}$ for average success rates is −0.37 at a testing noise level of 0.15. This result indicates that the success rate of the GFS with training noise level of 0.05 is more than the success rate of the GFS with training noise level of 0.15 by an amount of 0.37 at a testing noise level of 0.15. Therefore, from the results of minimum success rates and average success rates, the training noise level of 0.15 appears to be a judicious selection because this system also performs well with single types of measurement deltas like displacement- and moment-based measurement deltas.

Table 5.7 Effect of training noise level on average success rate at various noise levels for global damage detection using displacement, force, moments, and all measurements

α	Displacements		Forces		Moments		All	
	$\Delta S_{R0.05}$	$\Delta S_{R0.10}$	$\Delta S_{R0.05}$	$\Delta S_{R0.10}$	$\Delta S_{R0.05}$	$\Delta S_{R0.10}$	$\Delta S_{R0.05}$	$\Delta S_{R0.10}$
0.05	0.78	0.96	0	0.02	0	0	0	0
0.10	−0.35	0.59	−0.16	0.34	0.20	0.57	0	0.02
0.15	−0.25	2.18	−0.22	1.17	−0.19	0.28	−0.37	0.10

$\Delta S_{R0.05} =$ Difference between the average SR for training at $\alpha = 0.15$ and $\alpha = 0.05$
$\Delta S_{R0.10} =$ Difference between the average SR for training at $\alpha = 0.15$ and $\alpha = 0.10$
Negative value indicates better SR than the SR training at $\alpha = 0.15$

5.6.2.2 Local Damage Detection

The midpoints and standard deviations of the Gaussian membership functions for the rules for prediction of physical damage parameters and life consumption parameters in various parts of the blade using strain-based measurement deltas are tabulated in Table 5.8. These results are obtained at a training noise level of 0.05, which is found to be good for the strain-based measurement deltas. The midpoints in this table are normalized with the maximum value of the corresponding measurement deltas; corresponding values of the standard deviations are given in parentheses.

The success rates for prediction of physical damage parameters and life consumption parameters at various parts of the blades are tested at noise levels of 0.03, 0.05, and 0.10. Table 5.9 shows the success rate for all the rules with various noise levels. From Table 5.9, it is observed that the GFS for predicting the life consumption parameters in the matrix cracking zone in various parts of the blade using strain-based measurement deltas gives a success rate of 100% up to a noise level of 0.03. For the debonding/delamination zone, the GFS gives a success rate of 100% for a noise level of 0.03 except for two rules, Moderate D/D at Root and Severe D/D at Root. For a noise level of 0.05, the GFS gives an average success rate of 99.42% and a minimum success rate of 91.8%. The GFS gives a minimum success rate of 97.70% for the matrix crack zone and a minimum success rate of 91.8% for the debonding/delamination zone for a noise level of 0.05. It can be noted that the GFS gives a success rate of 100% for the debonding/delamination zone up to a noise level of 0.05, except for two rules, Moderate D/D at Root and Severe D/D at Root. However, for a higher noise level of 0.1, the system success rate starts falling rapidly for the rules which define matrix cracking towards the blade tip zone, and for the rules which define moderate and severe D/D at the root.

During the analysis of misclassification of the rules, the reason for the sudden fall in success rates at higher noise levels is that the GFS is unable to isolate the damage levels due to higher noise in the data. However, it is observed that even at the higher noise levels the GFS can isolate damage locations accurately.

Table 5.8 Midpoints and standard deviations of rules for local damage detection using strain-based measurement deltas

No.	Rule	$\Delta\varepsilon_{\text{Tip}}$	$\Delta\varepsilon_{\text{Outboard}}$	$\Delta\varepsilon_{\text{Center}}$	$\Delta\varepsilon_{\text{Inboard}}$	$\Delta\varepsilon_{\text{Root}}$
1	Undamaged	0.00 (0.18)	0.00 (0.14)	0.00 (0.17)	0.00 (0.13)	0.00 (0.15)
2	Small CD at Tip	0.08 (0.13)	−0.00 (0.16)	0.00 (0.10)	0.00 (0.13)	−0.00 (0.13)
3	Small CD at Outboard	0.00 (0.19)	0.02 (0.11)	0.01 (0.14)	0.00 (0.16)	−0.00 (0.16)
4	Small CD at Center	0.00 (0.16)	0.01 (0.10)	0.05 (0.15)	0.01 (0.16)	−0.00 (0.16)
5	Small CD at Inboard	0.01 (0.13)	0.01 (0.14)	0.01 (0.19)	0.04 (0.13)	−0.01 (0.12)
6	Small CD at Root	0.01 (0.14)	0.01 (0.12)	0.02 (0.13)	−0.00 (0.13)	0.05 (0.18)
7	Moderate CD at Tip	0.17 (0.12)	−0.00 (0.17)	0.00 (0.18)	0.00 (0.16)	−0.00 (0.15)
8	Moderate CD at Outboard	0.01 (0.11)	0.05 (0.13)	0.02 (0.20)	0.01 (0.19)	0.00 (0.19)
9	Moderate CD at Center	0.02 (0.10)	0.01 (0.18)	0.11 (0.12)	0.02 (0.12)	−0.00 (0.19)
10	Moderate CD at Inboard	0.04 (0.13)	0.02 (0.18)	0.04 (0.14)	0.09 (0.12)	−0.01 (0.18)
11	Moderate CD at Root	0.07 (0.19)	0.03 (0.13)	0.06 (0.18)	0.02 (0.14)	0.13 (0.16)
12	High CD at Tip	0.19 (0.20)	−0.00 (0.11)	0.00 (0.14)	0.01 (0.17)	−0.00 (0.17)
13	High CD at Outboard	0.01 (0.17)	0.05 (0.19)	0.02 (0.15)	0.01 (0.17)	0.00 (0.15)
14	High CD at Center	0.02 (0.11)	0.02 (0.11)	0.13 (0.15)	0.02 (0.19)	−0.00 (0.20)
15	High CD at Inboard	0.05 (0.17)	0.02 (0.12)	0.04 (0.20)	0.11 (0.17)	−0.01 (0.18)
16	High CD at Root	0.09 (0.13)	0.03 (0.18)	0.07 (0.12)	0.02 (0.11)	0.16 (0.11)
17	Very High CD at Tip	0.21 (0.17)	−0.00 (0.18)	0.00 (0.17)	0.01 (0.19)	−0.00 (0.12)
18	Very High CD at Outboard	0.02 (0.14)	0.06 (0.14)	0.02 (0.14)	0.01 (0.17)	0.00 (0.12)
19	Very High CD at Center	0.03 (0.14)	0.02 (0.19)	0.15 (0.17)	0.02 (0.18)	−0.00 (0.13)
20	Very High CD at Inboard	0.07 (0.15)	0.03 (0.19)	0.05 (0.17)	0.13 (0.14)	−0.01 (0.14)
21	Very High CD at Root	0.11 (0.17)	0.04 (0.14)	0.08 (0.14)	0.03 (0.19)	0.19 (0.17)
22	Slight D/D at Tip	0.41 (0.16)	−0.01 (0.15)	0.00 (0.16)	0.01 (0.14)	−0.00 (0.11)
23	Slight D/D at Outboard	0.04 (0.16)	0.12 (0.15)	0.03 (0.10)	0.02 (0.15)	0.00 (0.10)
24	Slight D/D at Center	0.10 (0.12)	0.04 (0.13)	0.33 (0.16)	0.05 (0.14)	0.01 (0.11)
25	Slight D/D at Inboard	0.28 (0.17)	0.09 (0.16)	0.18 (0.19)	0.36 (0.18)	0.06 (0.16)
26	Slight D/D at Root	0.48 (0.14)	0.15 (0.19)	0.31 (0.12)	0.18 (0.15)	0.62 (0.16)
27	Moderate D/D at Tip	0.59 (0.19)	−0.01 (0.20)	0.00 (0.12)	0.01 (0.17)	−0.00 (0.18)
28	Moderate D/D at Outboard	0.07 (0.20)	0.18 (0.11)	0.05 (0.17)	0.03 (0.10)	0.01 (0.18)
29	Moderate D/D at Center	0.23 (0.15)	0.08 (0.12)	0.59 (0.15)	0.09 (0.10)	0.04 (0.13)
30	Moderate D/D at Inboard	0.42 (0.20)	0.13 (0.20)	0.27 (0.19)	0.51 (0.12)	0.12 (0.19)
31	Moderate D/D at Root	0.83 (0.15)	0.24 (0.18)	0.48 (0.15)	0.31 (0.18)	0.94 (0.12)
32	Severe D/D at Tip	0.76 (0.13)	−0.01 (0.11)	0.00 (0.11)	0.02 (0.16)	−0.01 (0.11)
33	Severe D/D at Outboard	0.64 (0.17)	1.00 (0.19)	0.27 (0.13)	0.19 (0.15)	0.15 (0.12)
34	Severe D/D at Center	0.48 (0.11)	0.15 (0.17)	1.00 (0.15)	0.18 (0.15)	0.12 (0.12)
35	Severe D/D at Inboard	1.00 (0.18)	0.28 (0.15)	0.54 (0.14)	1.00 (0.20)	0.40 (0.10)
36	Severe D/D at Root	0.88 (0.14)	0.26 (0.19)	0.51 (0.13)	0.33 (0.14)	1.00 (0.17)

Table 5.9 Success rate of various rules for local damage detection using strain-based measurement deltas

No.	Rule	$S_{R0.03}$	$S_{R0.05}$	$S_{R0.10}$
1	Undamaged	100.00	100.00	100.00
2	Small CD at Tip	100.00	100.00	100.00
3	Small CD at Outboard	100.00	100.00	100.00
4	Small CD at Center	100.00	100.00	100.00
5	Small CD at Inboard	100.00	100.00	100.00
6	Small CD at Root	100.00	100.00	100.00
7	Moderate CD at Tip	100.00	98.60	81.10
8	Moderate CD at Outboard	100.00	99.60	81.50
9	Moderate CD at Center	100.00	99.90	86.00
10	Moderate CD at Inboard	100.00	100.00	95.50
11	Moderate CD at Root	100.00	100.00	99.90
12	High CD at Tip	100.00	99.90	47.20
13	High CD at Outboard	100.00	100.00	84.20
14	High CD at Center	100.00	99.80	68.40
15	High CD at Inboard	100.00	100.00	96.10
16	High CD at Root	100.00	100.00	83.80
17	Very High CD at Tip	100.00	99.30	84.10
18	Very High CD at Outboard	100.00	97.70	69.20
19	Very High CD at Center	100.00	100.00	89.90
20	Very High CD at Inboard	100.00	100.00	93.50
21	Very High CD at Root	100.00	100.00	99.40
22	Slight D/D at Tip	100.00	100.00	100.00
23	Slight D/D at Outboard	100.00	100.00	100.00
24	Slight D/D at Center	100.00	100.00	100.00
25	Slight D/D at Inboard	100.00	100.00	100.00
26	Slight D/D at Root	100.00	100.00	100.00
27	Moderate D/D at Tip	100.00	100.00	100.00
28	Moderate D/D at Outboard	100.00	100.00	100.00
29	Moderate D/D at Center	100.00	100.00	100.00
30	Moderate D/D at Inboard	100.00	100.00	100.00
31	Moderate D/D at Root	99.90	92.90	74.90
32	Severe D/D at Tip	100.00	100.00	86.00
33	Severe D/D at Outboard	100.00	100.00	100.00
34	Severe D/D at Center	100.00	100.00	100.00
35	Severe D/D at Inboard	100.00	100.00	100.00
36	Severe D/D at Root	99.90	91.80	73.90
	Avg.	99.99	99.43	91.52
	Min.	99.90	91.80	47.20

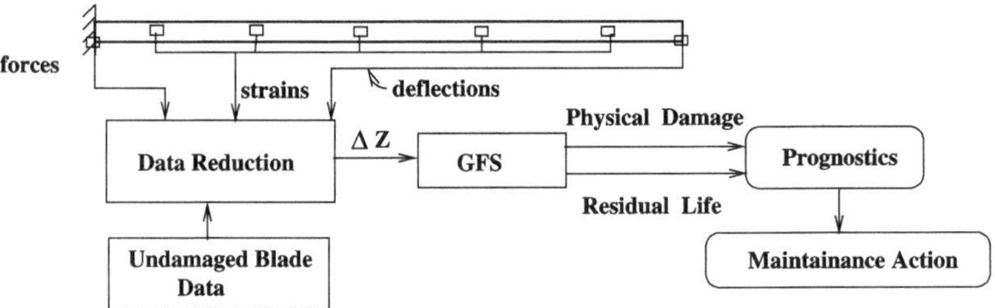

Fig. 5.21 Schematic representation of implementation of SHM

5.6.3 Implementation of the SHM System

The SHM system can be implemented on the helicopter rotor blade to predict the physical damage and residual life of the blade. The schematic diagram for implementation of the SHM system is shown in Fig. 5.21. As shown in the figure, for the global GFS, tip deflection and root forces, and for the local GFS, the strains measured at five locations can be compared with the database of measurements from the undamaged blade. Further, using the data reduction algorithm, the measurement deltas can be calculated and input to the GFS for predicting the physical damage and residual life of the blade. As shown in Table 5.2 for global SHM and in Table 5.3 for local SHM, maintenance norms can be developed to provide direct instructions to maintenance engineers. This will be helpful for optimal and safe use of the composite rotor blade as compared to the "Safe-life" method. As this method gives intermediate residual life zones, it will help in reducing the overhauling time.

This chapter showcases a comprehensive case study detailing the development of an SHM system for a composite rotor. It is clear that the development requires a knowledge of the system under consideration as well as a knowledge of the GFS [13–16]. This multidisciplinary nature is typical of realistic SHM problems.

5.7 Summary

In this chapter, an online SHM system is developed for predicting the global and local physical damage and life consumption of a composite rotor blade using the genetic fuzzy system. Hamilton's principle is used to derive the governing partial differential equations which are then solved using finite elements in space, normal mode transformation, and finite elements in time. The rotor and hub loads are obtained, and the vehicle trim equations are also solved simultaneously with the blade response equations. The mathematical model of the rotor blade is adjusted to include composite materials by adding new strain energy terms. A progressive damage growth model for composite materials which includes matrix cracking, delamination/debonding, and fiber breakage is integrated into an aeroelastic analysis for composite rotor blades. The aim of this modeling exercise is to develop a mathematical model for a damaged composite rotor blade. A phenomenological damage model is used to link the physics-based damage modes with the life consumption of the blade. The genetic fuzzy system is developed to maximize the success rate for damage detection. The genetic fuzzy system is able to determine the different types of damage, the location, and possible prognostic action required. The genetic fuzzy system is then tested for various measurements deltas with various noise levels. The effect of training noise on the genetic fuzzy system is also investigated. The genetic fuzzy system works quite well with noisy data and is recommended for online SHM of composite helicopter rotor blades. Implementation of the SHM on a composite rotor blade and its use in maintenance are discussed.

References

1. Hodges, D.H., Dowell, E.H.: Nonlinear equations of motion for the elastic bending and torsion of twisted nonuniform blades. NASA, TND-7818 (1974)
2. Leishman, J.G.: Modeling of subsonic unsteady aerodynamics for rotary wing applications. J. Am. Helicopter Soc. **35**(1), 28–35 (1990)
3. Chopra, I., Sivaneri, N.T.: Aeroelastic stability of rotor blades using finite element analysis. NASA CR 166389 (1982)
4. Bir, G., Chopra, I.: University of Maryland Advanced Rotorcraft Code (UMARC) theory manual. UM-AERO Report 92-02 (1992)
5. Ganguli, R.: Optimum design of a low vibration helicopter rotor using aeroelastic analysis and response surface methods. J. Sound Vib. **258**(2), 327–344 (2002)
6. Johnson, W.: Helicopter Theory. Princeton University Press, Princeton (1980)
7. Lim, J.W., Chopra, I.: Aeroelastic optimization of a helicopter rotor using an efficient sensitivity analysis. J. Aircr. **28**(1), 29–37 (1991)
8. Chandra, R., Chopra, I.: Structural response of composite beams and blades with elastic couplings. Compos. Eng. **2**(5–6), 347–374 (1992)
9. Adolfsson, E., Gudmundson, P.: Thermoelastic properties in combined bending and extension of thin composite laminates with transverse matrix cracks. Int. J. Solids Struct. **34**(16), 2035–2060 (1997)
10. Gudmundson, P., Zang, W.: An analytic model for thermoelastic properties of composite laminates containing transverse matrix cracks. Int. J. Solids Struct. **30**(23), 3211–3231 (1993)
11. Shahid, I., Chang, F.K.: An accumulative damage model for tensile and shear failures of laminated composite plates. J. Compos. Mater. **29**(7), 926–981 (1995)
12. Mao, H., Mahadevan, S.: Fatigue damage modelling of composite materials. Compos. Struct. **58**(4), 405–410 (2001)
13. Pawar, P.M., Ganguli, R.: Modeling multi-layer matrix cracking in thin walled composite helicopter rotor blades. J. Am. Helicopter Soc. **50**(3), 354–366 (2005)
14. Pawar, P.M., Ganguli, R.: Modeling progressive damage accumulation in thin walled composite beams for rotor blade applications. Compos. Sci. Technol. **66**(13), 2337–2349 (2006)
15. Pawar, P.M., Ganguli, R.: On the effect of progressive damage on composite helicopter rotor system behavior. Compos. Struct. **78**, 410–423 (2007)
16. Pawar, P.M., Ganguli, R.: Helicopter rotor health monitoring—a review. Journal of Aerospace Engineering **221**(5), 631–647 (2007). Proceedings of the Institution of Mechanical Engineers

Appendix
MATLAB Codes

MATLAB codes are available from http://extras.springer.com. A brief discussion of the codes is provided here.

1. *Code for calculating effective Young's modulus for given matrix cracking*: Matrix cracking effects are embedded in composite layups of cross-ply composite. This code is based on the theory given in Sect. 4.1.1. Material properties are given in the code. For a given level of matrix cracking, this code calculates the effective Young's modulus which will be used in a subsequent code for calculating variation in natural frequencies for given matrix cracking.

   File name: *matrix_cracking_code.m*

2. *Code for calculating change in frequencies at different crack densities*: This code calculates changes in the first eight modes of frequencies due to matrix cracking at ten different locations. This code is based on the theory discussed in Sects. 4.1.2, 3.1, and 4.2. For this code, the input file is "model2.inp" and output is stored in "model2.out". Inputs given in the input files can be understood by the comments given in the code. The inputs are in the sequence of Number of nodes, Number of elements, Number of material properties, Coordinates of the nodes, Connectivity, For each element Material index, moment of inertia, cross-sectional area and Mass Density of beam material. Effective material properties obtained from previous code for different crack densities are given in "mat1".

   File name: *frequency_calculation_code.m*
   Input file: model2.inp
   Output file: model2.out

3. *Code for training the genetic fuzzy system*: The first eight modes of frequencies for three damage levels at ten different locations are the input to this code. These numerical values are considered as midpoints of Gaussian membership functions, and the code calculates the values of standard deviations of the membership function for maximization of the success rate of each rule for noisy data of that rule. This code is based on the theory discussed in Sects. 4.3.1 and 2.3.

P.M. Pawar, R. Ganguli, *Structural Health Monitoring Using Genetic Fuzzy Systems*,
DOI 10.1007/978-0-85729-907-9, © Springer-Verlag London Limited 2011

File name: *Training_code_for_GFS.m*
Input file: *rul309.dat*
Output file: *tune.dat*

4. *Code for testing the genetic fuzzy system*: This code uses the frequencies developed from the code in point 2 and standard deviations obtained in the previous code to obtain the success rate of each rule for various noise levels. This code is based on the theory discussed in Sects. 4.3.2 and 2.3.

   File name: *Testing_code_for_GFS.m*
   Input file: *rul309.dat* and *tune.dat*
   Output file: *f09_415.dat*

Index

A
Aeroelastic analysis, 86
Analysis of misclassification, 81

C
Composite rotor blade, 92, 99
Cross-sectional stiffness, 100
Crossover, 34

D
Damage, 2
Damage detection, 2, 75
Damage modeling, 42
Data reduction, 97
Debonding/delamination, 95
Defuzzification, 30
Development of GFS, 75, 111
Diagnostics, 2

E
Effective elastic modulus, 68
Encoding, 33

F
Failure, 3
Fiber breakage, 97
Fitness evaluation, 33
Formulation of GFS, 45
Fuzzification, 36
Fuzzy logic system, 16, 26

G
GA performance factors, 35
Genetic algorithm, 15, 31
Genetic fuzzy system, 36
Genetic operations, 34
Global damage detection, 113

H
Hollow circular cross section, 68
HUMS system, 3
Hybridized soft computing methods, 18

I
Implementation, 124

L
Life of the structure, 98, 110
Local damage detection, 9, 114, 121

M
Matrix crack saturation, 108
Matrix cracking, 68, 69, 93
Measurement delta sets, 115
Measurement deltas, 103
Membership functions, 29
Missing and faulty measurements, 52
Modal analysis, 72
Modal based methods, 8
Mode shape, 63
Model based SHM, 4
Monitoring, 3
Mutation, 35

N
Neural network, 11
Noise, 39
Nonuniform beam, 54
Number of measurements, 51

P
Prognostics, 2
Progressive damage accumulation, 93

P.M. Pawar, R. Ganguli, *Structural Health Monitoring Using Genetic Fuzzy Systems*,
DOI 10.1007/978-0-85729-907-9, © Springer-Verlag London Limited 2011

R
Reproduction, 34
Root loads, 105
Rotor blade, 60
Rule generation, 37

S
SHM, 3
SHM approaches, 3
Soft computing methods, 11
Static response, 101
Strains, 108
Success rate, 38

T
Testing of GFS, 48, 78, 115
Tip response, 103

U
Uniform beam, 47
Usage, 3

MIX
Papier aus verantwortungsvollen Quellen
Paper from responsible sources
FSC® C105338

If you have any concerns about our products,
you can contact us on
ProductSafety@springernature.com

In case Publisher is established outside the EU,
the EU authorized representative is:
Springer Nature Customer Service Center GmbH
Europaplatz 3, 69115 Heidelberg, Germany

Printed by Libri Plureos GmbH
in Hamburg, Germany